The Logical Structure of Kinds

The Logical Structure of Kinds

Eric Funkhouser

OXFORD
UNIVERSITY PRESS

OXFORD
UNIVERSITY PRESS

Great Clarendon Street, Oxford, OX2 6DP,
United Kingdom

Oxford University Press is a department of the University of Oxford. It furthers the University's objective of excellence in research, scholarship, and education by publishing worldwide. Oxford is a registered trade mark of Oxford University Press in the UK and in certain other countries

First Edition published in 2014

Published in the United States of America by Oxford University Press
198 Madison Avenue, New York, NY 10016, United States of America

British Library Cataloguing in Publication Data
Data available

Library of Congress Control Number: 2014933806

ISBN 978-0-19-871330-2

For Erin and Mallory—two of a kind

Acknowledgements

I came up with the fundamental idea that drives this book—conceiving of kinds as property spaces structured by determination dimensions—over a decade ago. At that time I was a graduate student at Syracuse University working on a dissertation on mental causation. The idea emerged as I developed an alternative to Stephen Yablo's account of determinables and determinates, which he employed in an attempt to salvage the causal relevance of the mental. My focus at that time was more on mental causation and the exclusion problem, and not so much on offering a general theory of properties or their realization. I would like to thank my dissertation advisor, Robert Van Gulick, as well as Tamar Gendler and Ted Sider for their wise direction and encouragement at that stage. I couldn't have had a better intellectual and social environment in which to develop as a philosopher, and I miss those Syracuse days tremendously.

Over time the project shifted away from causation and more towards a general theory of properties and kinds, with specific applications to determination and realization relations. This book has *significantly* improved from the highly detailed, encouraging, and insightful comments by two anonymous referees. I am extremely humbled and grateful that so much care was given to help me improve this work and get it to publication. It is uplifting to experience such support and devotion to a common cause from the philosophical community.

I would also like to thank *Noûs*, *Philosophical Studies*, and *Philosophy Compass* for permission to include modified versions of some of my previously published work on determination and realization.

This book was a long time coming, and it was more of a solitary affair than is much contemporary philosophy. I greatly appreciate the patience of my two young daughters while they tolerated (though, to be honest, they didn't have much of a choice) the eccentricities of their father and his writing habits. Escaping to the bathtub might not be the best parenting, but at least it provided the seclusion needed to finish this book.

Contents

1

Introduction

Every conceptual scheme, or way of thinking about things, involves a classification into kinds. Examples are easy to come by. Biology classifies animals as mammals or reptiles, mathematics classifies numbers as prime or composite, and ethics classifies actions as duties or prohibitions. These classifications are then used to describe situations, make generalizations, advance laws and principles, and the like. It is no accident that we classify things into such kinds. Rather, thinking in terms of classifications into kinds is essential to gaining understanding that goes beyond immediate perception and memory. Our understanding would be severely impoverished were we to think only in terms of particulars. John Locke, writing about general terms, put the point like this:

> . . . *a distinct Name for every particular Thing, would not be of any great use for the improvement of Knowledge*: which, though founded in particular Things, enlarges it self by general Views; to which, Things reduced into sorts, under general Names, are properly subservient.[1]

Given the central role that kinds and kind terms play in our understanding of the world, it is reasonable to think that we would benefit from a careful, logical investigation into the nature of kinds. Toward this end this book will address some logical and metaphysical issues underlying the taxonomies of the various conceptual schemes, with special attention given to scientific kinds. A logical investigation of kinds searches for the generic structure shared by taxonomies, their kinds, and the relations among them. Among other advantages, I believe that providing a theory of the logical structure of the kinds posited by the various conceptual schemes clears the way to a better understanding of how various conceptual schemes relate to one another. For example, to

[1] Locke (1975: 410).

mention a prominent example of long-standing interest to philosophers, we can thereby gain greater insight into how neuroscientific kinds relate to psychological kinds.

Such a logical theory of kinds is not developed simply for its own sake, but also for its applications to practical questions such as those concerning the methodology and autonomy of the "special sciences"—the sciences besides physics (basic science). In this book a logical theory of kinds is applied to some core problems and topics in contemporary philosophy of mind and science in particular. To anticipate some of the benefits of having such a theory, here are some examples:

kind identity and specification: What conditions must be met in order for distinct concepts or terms to pick out the same kind? Many taxonomies progress from quite general to more specific classifications, but what is the nature of the relationship between a more specific kind and the more general kind that it specifies (as when we say that *short-term memory*[2] specifies *memory*)?

multiple realizability: What must be the case in order for one kind to be multiply realized by other kinds? How are such kinds related to their multiple realizations? And what consequences do the answers to these questions have for reduction and autonomy?

reduction: Under what conditions do the kinds of one conceptual scheme reduce to (e.g., are identical to or "nothing over and above") the kinds of another conceptual scheme?

autonomy: Under what conditions is one conceptual scheme autonomous with respect to another conceptual scheme?

mental causation: What conditions must be met in order for a causal transaction to be distinctively mental (or, for that matter, chemical, biological, etc.)?

These are just a few of the salient applications to one active branch of current philosophy. The theory and its applications generalize to other sciences and conceptual schemes as well.

Much has been written about both, on the one hand, the metaphysics of properties and their kinds and, on the other hand, issues in the philosophy of mind such as realization, reduction, and mental causation. This book

[2] Words attempting to refer to kinds, kind terms, will be italicized throughout. Some other words will also be italicized for emphasis, but this should not cause any confusion.

bridges a gulf between the metaphysics literature and philosophy of mind literature. Contrary to many works in the philosophy of mind, I offer a very thorough metaphysical account of properties and their kinds. This is necessary given the prominent role that kinds and kind terms play in accounts of realization, reduction, mental causation, and the like. But contrary to many works on the metaphysics of properties and their kinds, my metaphysical account of properties and their kinds is largely driven by appreciating relations, like determination and realization, that many philosophers of mind have claimed hold among various psychological and neuroscientific kinds.

Though these are very modern-sounding terms and problems, I see the project developed here as part of a rich philosophical tradition that traces back, as does much of philosophy, to Socrates and Plato. Socrates was often interested in seeking the Form of some ethical concept, such as justice, virtue, or piety. I understand this Socratic project as the search for the commonalities in virtue of which different things all belong to the same kind. For example, in seeking the Form of virtue Socrates was searching for that which all virtuous acts or people have in common that makes them virtuous. Consider this classic exchange from Plato's *Meno*:

> SOCRATES: I seem to be in great luck, Meno; while I am looking for one virtue, I have found you to have a whole swarm of them. But Meno, to follow up the image of the swarms, if I were asking you what is the nature of bees, and you said that they are many and of all kinds, what would you answer if I asked you: "Do you mean that they are many and varied and different from one another in so far as they are bees? Or are they no different in that regard, but in some other respect, in their beauty, for example, or their size or in some other such way?" Tell me, what would you answer if thus questioned?
>
> MENO: I would say that they do not differ from one another in being bees.
>
> SOCRATES: If I went on to say: "Tell me, what is this very thing, Meno, in which they are all the same and do not differ from one another?" Would you be able to tell me?
>
> MENO: I would.
>
> SOCRATES: The same is true in the case of the virtues. Even if they are many and various, all of them have one and the same form which

> makes them virtues, and it is right to look to this when one is asked to make clear what virtue is. Or do you not understand what I mean?[3]

We can think of these Forms as providing us with the essences of their kinds. Importantly, Socrates thought that knowledge of these Forms or essences was a prerequisite to knowing many other truths about those kinds. In fact, he thought it was necessary for real knowledge at all.

Like Socrates, I think it is important to understand the Form, essence, nature, or structure of various kinds. But my pursuit of the structure of kinds differs from that of Socrates in three important ways. First, whereas Socrates was primarily interested in discovering the Forms of ethical concepts, I am more interested in discovering the structure of scientific concepts and the kinds that they pick out. This is simply a difference in focus, without any disagreement. Second, I am interested in the logical form of these kinds rather than looking for particular essences or concerning myself with their ontological standing (e.g., whether they are Platonic Forms that concrete particulars participate in). Third, I differ from Socrates in the method I endorse for discovering the structure of kinds. Whereas Socrates thought that knowledge of Forms is to be acquired by *a priori* reflection, I believe that such knowledge is typically *a posteriori*. This is especially true for scientific kinds. In this regard, I follow a popular trend in recent metaphysics. Since the pioneering work of Hilary Putnam (1975), David Armstrong (1978), and Saul Kripke (1980), it is now widely acknowledged that the essences of the kinds picked out by our kind terms are often largely determined by contingent facts about our world that can only be discovered by empirical investigation. I see myself and this work in particular as participating in this more recent tradition.

The original, ancestral work on the logic of kinds is perhaps Aristotle's *Categories*.[4] Among his ten categories, *qualification* comes closest to corresponding to the modern understanding of properties or, at least, the types of kinds of primary interest here. The notion of differentiae was central to Aristotle's conception of taxonomies, and the present logical investigation of kinds can be seen as a systematic treatment, applied to adjectival kinds, of a quite similar notion. Among other things, Aristotle was interested in

[3] Plato (1997: 872–873).

[4] Gaines (2009) drew my attention to the connection between my work on determination dimensions and Aristotle's classic work on taxonomies.

discovering the structure that explains when things differ as falling under a kind, and when they are of truly distinct kinds. He writes:

> The differentiae of genera which are different and not subordinate one to the other are themselves different in kind. For example, animal and knowledge: footed, winged, aquatic, two-footed are differentiae of animal, but none of these is a differentia of knowledge; one sort of knowledge does not differ from another by being two-footed. However, there is nothing to prevent genera subordinate one to the other from having the same differentiae. For the higher are predicated of the genera below them, so that all differentiae of the predicated genus will be differentiae of the subject also.[5]

In my terminology, to be expounded on shortly, *determination dimensions* occupy a role similar to Aristotle's differentiae. Some of the key points that we find Aristotle expressing in this passage also reappear (in some form) in my account of determination. One of my main claims is that this ancient distinction is critical to current debates over reduction and autonomy in philosophy of mind (and philosophy of science, more generally).

It has long been understood that a theory of kinds is particularly pressing for understanding foundational issues in the empirical sciences. The problems of induction also drive some of these concerns. The sciences are largely in the business of discovering inductive generalizations, and the notion of a kind—along with related concepts like resemblance or similarity—is central to inductive inference. As W. V. O. Quine reminds us:

> Two green emeralds are more similar than two grue ones would be if only one of the grue ones were green. Green things, or at least green emeralds, are a kind. A projectible predicate is one that is true of all and only the things of a kind. What makes Goodman's example a puzzle, however, is the dubious scientific standing of a general notion of similarity, or of kind.
>
> The dubiousness of this notion is itself a remarkable fact. For surely there is nothing more basic to thought and language than our sense of similarity; our sorting of things into kinds.[6]

This work, while not taking on the problems of induction, attempts to go some way in clearing up the "dubious scientific standing" of kinds. I also hope to shed some light on the status and logic of resemblance and similarity, issues core to scientific theorizing.

We can now see that we have many good reasons to investigate the logical structure of kinds. If nothing else, a quick look at select passages

[5] Aristotle (1984: 4). [6] Quine (1969: 116).

from the likes of Plato, Aristotle, Locke, and Quine should remind us of the long-standing philosophical importance of investigating the nature of kinds.

1.1 Scope

Let us start our investigation by considering an obvious yet interesting truth about human inquiry. Physicists, psychologists, artists, ethicists, and other inquisitive people can all have true and fruitful thoughts about the same phenomenon, all the while conceptualizing that phenomenon in radically different ways. A father playfully tosses his young daughter into the air. The physicist sees gravity pulling her down, the psychologist sees the daughter's trust grow, the artist sees the beautiful line of her trajectory, and the ethicist sees a father's duty fulfilled. These are all true and appropriate, though radically different, thoughts about the very same scene.

It is obvious that there are these different conceptual schemes—ways of thinking about things—that come with their distinctive vocabularies. As our example shows, these different conceptual schemes are not always at war or in conflict with one another, but they often can coexist with equal claim to the honor of truth. In this book I am interested in certain relations that sometimes hold across and within such conceptual schemes, the sciences in particular. The relations that are of particular interest to me are *necessitation* relations among kinds. My main goal in this book is to give an account of two such synchronic necessitation relations and illustrate their importance to scientific and metaphysical theorizing, as well as the importance of our *recognizing the differences* between these relations. I will argue that understanding the differences between these two necessitation relations is essential to adequately understanding and fruitfully employing scientific kind terms and the taxonomies to which they give rise.

So, what is an example of such a necessitation relation? Through reflection or empirical investigation we sometimes discover that whenever some description is true of an object or situation some other description must be true of it as well. Let us consider some examples. Whenever an object is crimson, it is also red. Whenever a person is in some particular neuroscientific state, she is undergoing conscious experiences. And whenever something is water, it is also H_2O. In each of these examples the condition specified by the former description necessitates that specified by the latter description. I will later argue that each of these three examples presents an interestingly different variety of necessitation—*determination, realization*, and *identity* respectively. Taking our second example, the neuroscientist and psychologist can each have true beliefs about the same person, framed in the distinctive vocabularies for their conceptual schemes. The neuroscientist believes that some neural pattern has been activated, say, and the psychologist believes that the patient is in a euphoric mood. Let us suppose that they are both correct. This case differs from the example discussed above involving the father and daughter in that the activation of that neural pattern is supposed to necessitate the euphoric mood. In contrast, it is plausible that no such necessitation relations hold for the descriptions given in our earlier example—e.g., not all instances of such beautiful trajectories are fulfillments of parental duties, and so too for any other combination of descriptions from that example. So, in this one example, we are supposing that there is an intimate connection between neuroscience and psychology.

Such necessitation relations establish bridges across conceptual schemes. Whenever such relations obtain, truths from one conceptual scheme guarantee truths in another conceptual scheme, as some neuroscientific truths guarantee particular psychological truths. It is natural then to wonder whether such conceptual schemes are truly independent or *autonomous*, at least to the extent that such relations hold. Autonomy comes in different varieties. The kinds posited by a conceptual scheme, CS_1, can be autonomous with respect to the kinds posited by another conceptual scheme, CS_2, in at least three different senses. First, they could be *ontologically autonomous* in that the kinds posited by CS_1 are not identical to (or would not be identical to, in cases of false theories) any of the kinds posited by CS_2. Second, they could be *explanatorily autonomous* in that the explanations offered in the vocabulary of CS_1 are not simply notational variants of explanations offered in the vocabulary of CS_2. And third, they could be *methodologically autonomous* in that the appropriate training, methodology, equipment, and standards for investigating the kinds posited by CS_1 are not inherited from or otherwise determined by the appropriate methodology for CS_2. So if CS_1 is methodologically autonomous with respect to CS_2, then the way in which CS_2 is to be practiced is not necessarily the way in which CS_1 is to be practiced. These three senses are not completely independent. For example, if CS_1 is ontologically autonomous with respect to CS_2 it is *prima facie* plausible that CS_1 is also explanatorily and methodologically autonomous with respect to CS_2. After all, the conceptual schemes are concerned with different aspects of the world. Among other things, in this book I am concerned with examining the ontological, explanatory, and methodological consequences that follow from such necessitation relations (or the lack thereof) across and within the sciences. So, these different senses of 'autonomy' will receive much greater attention later (Chapters 4 and 5).

The primary necessitation relation of interest in this book is the specification relation called *determination*. The treatment of determination is quite technical, but the big point at issue is also supposed to be quite commonsensical: under what conditions do we have merely a difference in specificity (or degree), and under what conditions do we have a genuine difference of kind? (Here, recall our passage from Aristotle.) The theory I develop to answer this question operates on two different levels. On one level, I argue for certain general claims concerning *taxonomies* in the sciences and other conceptual schemes, as well as for general claims concerning relations among these taxonomies. These claims are relevant to many issues within the philosophy of science, including the individuation of kinds, multiple realizability (MR), reduction, autonomy, and causation. I believe that most, if not all, of these claims can be accepted by those who spurn metaphysics or are simply non-committal on metaphysical issues. On another level, I develop a *metaphysical theory of properties* to fit with these claims concerning scientific taxonomies. But I believe that one can accept my claims on the taxonomic level while not accepting, or simply ignoring, the claims made at the metaphysical level.

There is much to be said about the structure of kinds and their instances, and, of course, not all that is worth saying is said here. So, I should clarify at the outset the limited focus of this book. I aim at articulating the general structure that is common to all kinds, as well as some quite general relations that these kinds can enter into with one another. I do not pursue questions concerning the essences of particular kinds except for illustrative purposes along the way to help explain the theory. It is important to recognize that kinds are quite varied. There are functional kinds (*digestion*), qualitative kinds (*pain*), mathematical kinds (*prime*), ethical kinds (*duty*), aesthetic kinds (*expressionism*), epistemic kinds (*justified belief*), artifactual kinds (*computer memory*), social kinds (*race*), and others. It would be foolish for me to defend a substantive thesis concerning how all such kinds should be individuated. However, I think there are certain formal features that are common to all kinds and their taxonomies. My main claim concerning the structure of kinds is that they have distinctive dimensions that partially capture their essences, and these dimensions determine the determinates of those kinds while also distinguishing them from their realizations (if they have any). The vocabulary here—determinates and realizations—is technical, but, in Chapter 2, I clarify these key terms and argue for this main claim.

1.2 Terminology and Assumptions

That said, let me make my relevant metaphysical assumptions explicit. First, in developing my metaphysical theory of properties I simply assume *Property Realism*. As I am using the term, a Property Realist is anyone who believes that properties are irreducible inhabitants of the world that are ontologically distinct from concrete particulars and sets thereof. Property Realists, then, are those who accept Platonic universals, Aristotelian universals, or abstract particulars (tropes). Second, I also assume *Particularism*. This is the thesis that everything that exists is particular. Accepting both of these claims makes me, in contemporary vocabulary, a trope theorist. Obviously, if I were developing a comprehensive metaphysics of properties I would need to argue for these basic assumptions. But such argumentation would distract from the central claims of this book which, I think, are less metaphysical and extremely important in their own right. There is metaphysics, and then there is metaphysics that is of interest only to the analytic metaphysician. Through an exercise of self-restraint, I have opted to stick with the former project. (Some, I am sure, will think that I have failed miserably!) I then offer a metaphysical theory of properties to the following extent: on the assumption that properties exist as particulars, I offer a theory of their generic structure as well as their relations to the kinds countenanced by the various conceptual schemes that they fall under.

This book presents a partial theory of property tokens and their types. Assuming knowledge of the type/token distinction, what does this mean? The division between a theory of tokens and a theory of types makes sense in other domains. For example, suppose we were interested in having a theory of cars. It would make sense to distinguish between theories of car tokens and theories of car types. A theory of car tokens explains the nature of individual cars—what they are *like*. A theory of car types, in contrast, deals with the varieties of cars, the ways of taxonomizing them, and the relations among such varieties.

I hope that distinction makes good sense to you with respect to cars. But what about properties? Some philosophers might think that there is something fishy about marking a distinction between theories of property types and property tokens. The reason for their unease might be that they think *properties just are types*. A theory of car tokens explains the nature of particular cars, but properties are not particulars, the claim goes, so there can be no theory of property tokens. In contrast with properties conceived of as particulars, these philosophers hold that properties are universals.

At this stage it will be helpful to explicitly define some of these critical terms. The words 'particular' and 'universal' are used in various ways by various philosophers, but I will use them in the following senses:

particular: that which is incapable of being wholly present at distinct spatial locations at a time.
universal: that which is capable of being wholly present at distinct spatial locations at a time.[7]

The distinction marked between these terms has occasionally been conflated with the distinction between the concrete and abstract. But the latter terms should be understood, for present purposes at least, as follows:

concrete: a gross and complete object, such as the total occupant of some region of space-time.
abstract: some feature or aspect of an object, existing in its own right in addition to (though perhaps dependent on) the object to which it belongs.

Note that in this sense 'abstract' refers to mind-independent features of the world. Of course, 'abstract' is used in various other ways. Sometimes it is used to refer to non-spatio-temporal entities—e.g., numbers—and other times it is used as a verb referring to the process of mentally isolating or subtracting some feature from its object.[8] The present usage

[7] Some have rejected the legitimacy of this distinction. See, for example, MacBride (2005). To be clear, I am making this distinction in order to reject universals as understood in the robust sense of the Platonist or Aristotelian about properties.

[8] The latter usage can be found in Locke (1975: II.xi.9).

contrasts with both of these alternatives. First, 'abstract' will be used to refer to features of objects that can be either spatio-temporal or non-spatio-temporal. Second, these features are supposed to be the (typically) mind-independent entities that can correspond to our mental acts of abstraction. For example, the mass of this apple is abstract, and I can conceptualize the apple's mass through the mental act of abstraction. But, again, I will not be using 'abstract' to refer to such mental acts.

The above distinction between these two dichotomies—particular/universal and concrete/abstract—mirrors that made by D. C. Williams (1931).[9] Like Williams, I will utilize these two distinctions to explain the property theory that I favor. Again, I am assuming Particularism, so I reject properties as universals. I also assume Property Realism, so I accept the reality of abstractions. In making these assumptions, I allow for a distinction between theories of abstraction tokens and theories of abstraction types. In order to avoid the cumbersome terminology 'abstraction tokens' and 'abstraction types,' I will utilize the above distinctions to define 'property' and 'kind':

> **property**: an abstract particular (i.e., trope), for which a proper noun is appropriate.
>
> **(property or adjectival) kind**: a type (category, variety, class, etc.) of property, for which a common noun is appropriate.

Of course I recognize that 'property' is understood in other ways by different theorists. In particular, many use 'property' to refer to what I am calling 'kinds.' But I do not think that my usage is idiosyncratic. There is nothing strange, to my ear at least, about saying that a given property of a car, its mass, say, caused the old, wooden bridge to break. And here we are using 'property' to pick out a particular—the mass of this car as opposed to that of equally massive cars (i.e., cars of the same *mass* kind).

Likewise, the use of 'kind' to refer to a type should be unobjectionable as these are near synonyms. I intend for the term to be used liberally, so

[9] The concrete/abstract distinction is a bit more difficult to characterize, I believe, than the particular/universal distinction. It might be helpful to recall D. C. Williams's (1931) attempt at marking the concrete/abstract distinction: "A concrete entity is one which affords or can afford the total content of a spatio-temporal volume, or of a chunk. An abstract entity is one which does not and cannot afford the total content of a spatio-temporal volume, or chunk" (588). This definition is good for the most part, but, unfortunately I think, it rules out the possibility of concrete objects that are non-spatio-temporal.

that it is an open possibility (at the outset of investigation, at least) that any old classification of properties forms some kind. So, 'kind' should not be understood as shorthand for 'natural kind.' The restriction of 'kinds' to refer only to types of *properties*, though, might be objectionable. After all, aren't there kinds of *objects* (*concreta*)—e.g., tigers and rocks? Of course there are such kinds—what I will call *substantival kinds*—in the common and appropriate use of the term. So, my unqualified use of the term 'kind' should typically be understood as shorthand for 'property kind' or 'adjectival kind.' The terms 'property kind' and 'adjectival kind' simply seem too cumbersome, so I have dropped these qualifications except when I think that the qualification needs to be emphasized.

It will prove worthwhile to point out and illustrate the distinction between property kinds (adjectival kinds) such as *mass* and object kinds (substantival kinds) such as *tiger*. The distinction here is metaphysical rather than grammatical. For terms that appear as grammatical subjects of sentences can nevertheless refer to adjectival kinds—e.g., '*Red* is a *color*.' Though I am primarily concerned with offering an account of adjectival kinds, the theory I develop can also be extended, with some slight alterations, to substantival kinds. This point is explicitly addressed in §2.8, where the necessary alterations are also made.

I ask the reader to bear in mind that the six words defined above will be used in these technical senses throughout. These definitions correspond to the stated scope of this book. To the extent that this book details interesting logical and metaphysical relations that can hold among kinds, offers suggestions for taxonomization, and insights on MR and reduction, a type-level theory of properties is presented. And to the extent that this book offers a metaphysical theory of the nature of properties and their individuation conditions, a token-level theory is also presented. These combine to form a formidable, if not comprehensive, theory of properties as particulars.

1.3 Relevance and Importance

There is intrinsic value in possessing an accurate and informative theory of properties, kinds, and the taxonomies to which they give rise. But I have also developed my views here with an eye directed at applications to contemporary debates over reduction, autonomy, and mental causation. Take mental causation, in particular. The primary problem of mental causation, as I understand it, is whether mental causation occurs and, if so, how it occurs. The major step to resolving the problem of mental causation, if it is to be resolved at all, is to first resolve more general issues from metaphysics and the philosophy of science that are not unique to mentality. These include: (1) developing an account of properties, events, and/or the other ontological categories that are relevant to the causal relation, (2) developing an account of causation that explains both the relation and the nature of its *relata*, and (3) developing an account of reduction and autonomy. With these in place, the metaphysician or philosopher of science can then turn the keys over to philosophers of mind, cognitive scientists, neuroscientists, etc., to resolve the empirical and conceptual concerns that are specific to mentality. These include issues like determining if the mental is related to the physical in the manner required for reduction, discovering whether mental properties or events enter into the relations that are required for causation, and discovering the nature of mental properties or events.

This book offers a partial theory of properties and their kinds that addresses, to some extent at least, each of these three issues. First, I develop a theory of how properties relate to their kinds and how both are to be individuated. This theory is motivated by a distinction between two types of synchronic necessitation relations that can hold among properties and their kinds—determination and realization. I offer no account of the causal relation itself. However, it is widely accepted that properties play an essential role in causal transactions. So to the extent that I provide individuation conditions for properties, I am likely contributing to the individuation of the causal *relata* as well. Finally, the concepts of determination and realization are employed to articulate and argue for various senses of 'reduction' and 'autonomy.' I reach no conclusions regarding the actuality and manner of mental causation. Instead, I aim to go some way towards resolving the foundational issues in metaphysics and philosophy of science preparatory to such answers.

1.4 Methodology

The core of this theory of properties and their kinds is presented in Chapter 2, with later chapters responding to objections, providing applications, and illustrating by examples. The method here is to construct a theory of properties by considering the data for which a good property theory should account. In developing this theory, I give particular weight and attention to synchronic necessitation relations that can hold among properties and their kinds. I will not provide anything resembling a deductive argument for the conclusions I reach. Instead, I assert various claims that I believe have intuitive plausibility and which, I hope to illustrate, account for the data to be explained. Prominent among these is a logical or metaphysical distinction between the distinct relations *specification* and *realization* that can hold among kinds. This distinction is critical to the resulting theory. The theory is then further justified by the fruit it bears when applied to issues in other domains—e.g., MR, reduction, and causation. Chapter 2 gives special attention to the specification relation, and I argue that adjectival kinds specify one another in virtue of being related as determinate to determinable. I then offer an analysis of the determination relation. This may seem like an idiosyncratic aspect of kinds. But to the contrary, I hold the determination relation to be central to understanding the structure of kinds, and I even apply this analysis of the determination relation to provide individuation conditions for properties and their kinds.

I will normally write the original formulations of my theses concerning property or kind individuation, MR, reduction, etc., in terms that assume Property Realism. However, as I think that the insights hold regardless of this assumption, I will also try to add a version of each thesis that is friendly to the Nominalist. This is consistent with my aim of addressing both the metaphysician and non-metaphysician. I do not think that my most fundamental claims are metaphysical in a robust or pejorative sense. Rather, I see them as having a broad appeal to our understanding of the various sciences, their relations to one another, and their autonomy.

In Chapter 3, I present and respond to various objections one could reasonably raise against this property theory. Prominent among these objections are challenges to my claim that kinds possess unique determination dimensions, challenges to the very existence of determinables, and inquiries concerning the ontological priority that should be given to kinds or properties.

Chapter 4 focuses on the concept of realization and the critical role that MR plays in arguments for various forms of non-reductionism and autonomy. In this chapter I argue for an account of realization that differentiates it from the determination/specification relation discussed earlier. This chapter concludes with objections to alternative accounts of MR.

Chapter 5 continues the topic of MR, and it is here that I offer my positive account of MR. This theory is defended from some anticipated objections, and I also show how it generates the theoretical results that we should expect of a satisfactory theory of MR. I attempt to show, in some detail, how this theory of determination and realization applies to actual scientific examples. Here we see the relevance of the taxonomic and metaphysical theory to "real-world" examples. This chapter then concludes with some more speculative claims concerning MR and scientific reduction.

2

Determination and Kinds

This chapter provides the core of my theory of the logical or metaphysical structure of properties and their kinds. Central to this theory is a distinction between *specification* and *realization*, two distinct relations that kinds can bear to one another. This is a logical or metaphysical distinction that has great importance in understanding the relations between our various systems of classification, even if the more metaphysical points that I develop prove to be misguided or insignificant. In other words, this theory is relevant even for those—philosophers of mind and science, in particular—who are skeptical of analytic metaphysics. Later chapters apply this theory to the topics of MR, reduction and autonomy, and "real-world" examples. The account offered in this chapter is rather technical at times, but the details will later prove essential to understanding solutions to interesting philosophical problems in other domains.

2.1 Kind Necessitation

I want to begin this chapter by motivating the need for a theory of properties and their kinds, and the necessitation relations among them in particular. To get a sense of the use to which such relations are employed, consider the following philosophical topics in which synchronic necessitation relations among kinds are prominent:

1. Physicalism: At issue here is whether the physical truths determine all the other contingent truths, in particular the psychological truths, of our world. Various necessitation relations—e.g., conceptual entailment, global supervenience, strong supervenience, and realization—have been invoked to formulate the thesis of physicalism. The question of physicalism comes in two parts. First, what relations must hold between physical kinds and other kinds in order

for physicalism to be true? And second, do these relations actually obtain?

2. Higher-Level Causation: Whether or not physicalism is true, there might be necessitation relations, such as supervenience, that hold among the kinds posited by physics and various higher-level kinds. If physics is also causally closed, then this could threaten the possibility of novel causation among instances of higher-level kinds.
3. Reduction: At issue here is the relation that must hold among the kinds posited by two conceptual schemes in order for the kinds of one conceptual scheme to reduce to the kinds of the other.
4. Taxonomies: Necessitation relations amongst kinds can also help shape the taxonomies of conceptual schemes (in science, ethics, etc.). Taxonomies typically include kinds ranging from the general to the more specific, with the more specific necessitating the more general. If we are interested in how best to taxonomize, we might then be interested in discovering the specific types of necessitation relations that are relevant, in general, to this process. It would be helpful to know when the discovery of a novel necessitation relation involving some kind *K* of taxonomy *T* should alter or be included in the structure of *T* and when such a discovery is irrelevant to its structure.
5. Inheriting Truths: Discovering necessitation relations can also be a way of discovering new truths about a kind, because this is a way of inheriting truth. For example, suppose that all instances of kind K_1 are also instances of kind K_2, because K_1 necessitates K_2. To give a concrete example, suppose that all promises are perfect duties. By discovering this necessitation relation among moral kinds, we conclude that it is always morally unacceptable to break a promise. The moral unacceptability here is inherited from the moral unacceptability of violating any perfect duty, because promises are a kind of perfect duty and it is always morally unacceptable to violate a perfect duty.
6. Mechanistic Explanation: Other times necessitation relations among kinds reveal mechanistic relationships as opposed to varieties of a kind. For example, kinds K_1–K_n might collectively necessitate kind K_{n+1} even though kinds K_1–K_n do not constitute a distinctive variety of K_{n+1}. To give an actual instance, it might be that various sonic kinds (*wavelength, amplitude*, etc.) of particular values necessitate a certain

aesthetic kind (e.g., *sad music*). Though these sonic kinds realize the aesthetic kind, it does not follow and (as I will argue later) it is not the case that this combination of sonic kinds is a variety of *sad music*. In general, it is not the case that every distinct type of mechanism for a kind forms a distinct variety of that kind. This is the topic of MR, which is much discussed in Chapters 4 and 5.

Each of these topics provides us with good reasons to be interested in understanding kind necessitation relations, and more examples and applications could be given. Understanding kind necessitation is then critical to various scientific, metaphysical, ethical, aesthetic, and other enterprises. So let's begin our investigation by asking what it even means for one kind to *necessitate* another kind. Kinds can stand in many interesting logical, conceptual, metaphysical, and nomic relations to one another. Necessitation is one such relation, and it comes in all of these modal varieties. Let's understand kind necessitation as follows:

Kind necessitation: Kind *A* necessitates kind *B* if, and only if, whenever an object is of kind *A* it is also of kind *B*.

A few comments will help to clarify and justify this simple definition.

First, kind necessitation can be synchronic, diachronic, or non-temporal. Supervenience and realization are examples of synchronic kind necessitation relations. Causation is the prime example of a (typically) diachronic kind necessitation relation. Certain kinds of conceptual entailments are non-temporal. My primary interest here is with synchronic kind necessitation relations and their relevance to taxonomies in the sciences and elsewhere.

Second, the modal force of this necessitation can be read in different ways and with different strengths. As mentioned above, the necessitation may be logical, conceptual, metaphysical, physical, etc. The 'whenever' aspect of this definition can be read or qualified in any of these different senses. So, the above is actually a definition schema for various kind necessitation relations. We can consider more precise relations, as follows:

Logical kind necessitation: e.g., *red and square* necessitates *square*
Conceptual kind necessitation: e.g., *knowledge that p* necessitates *p*
Metaphysical kind necessitation: e.g., *water* necessitates H_2O
Physical kind necessitation: e.g., *conscious* necessitates *physical*

Some of these examples are debatable, but I will not worry about defending them here. Further, I am not sure how each of these modalities, strictly speaking, is to be understood. The examples are offered simply to portray a sense of the varieties of kind necessitation. Here I also wish to note that there will be some disagreement over the modal force accompanying the synchronic kind necessitation relations of present interest, such as realization.

Third, the intimacy of the connection between kinds related by kind necessitation can vary for different pairings. This variance will often correspond to the modal distinctions just marked, but the distinctions also cross-cut one another. Sometimes a kind necessitates another kind in a very trivial sense, as when every object falls under the latter kind. For example, *green* kind-necessitates *wet or not wet* even though there is no intimate connection between these distinct kinds. There is no intimacy here, because any other kind term could be substituted in for '*green*' with truth being preserved. Other times the connection is more interesting and intimate because not every kind necessitates the latter kind, but the kind necessitation could still be, say, accidental or due to a common cause. This is the case, for example, when a symptom kind-necessitates a disease, though the symptom in no way explains or brings about the disease. We reach real intimacy of kind necessitation when the occurrence of one kind *explains* the occurrence of the other. In such cases we say that an object is of one kind *because* or *in virtue of* the fact that it is of some other kind. For example, something might be red in virtue of being crimson. Or someone might be in pain because he is in central nervous system state 1204. This explanatory language, "in virtue of" and "because," is difficult to cash out and is typically left at an intuitive or metaphorical level of understanding. But I think that some progress can be made, as evidenced in the remainder of this chapter. Finally, there is the most intimate kind necessitation relation of all—identity. The kind necessitation relations of present interest all fall on the intimate end of the spectrum, with an emphasis on those that are asymmetric and synchronic.

These "in virtue of" locutions have been taken with metaphysical seriousness. In fact, there has been a recent move to lessen the role of modal concepts (e.g., necessitation) in metaphysics and to replace them with an arguably better-suited notion of ontological priority, grounding, or fundamentality.[1] They very well might be correct that metaphysics as practiced

[1] The recent literature here is large. But, for prominent examples see Fine (1994), Rosen (2010), and Sider (2011).

over the last fifty years or so has overextended the applicability of modal concepts. But I think that the present work's emphasis on necessitation as a key concept in articulating determination and realization is appropriate. This is because I aim to give an account of empirically discoverable relations that are investigated by the sciences and other empirical and practical disciplines. However sympathetic I may be to the "in virtue of" locutions, more fine-grained and esoteric (e.g., empirically inaccessible) relations are ill-suited for this task.

2.2 Specification and Realization

Let's consider some specific examples of kind necessitation, with the ulterior motive of marking a critical distinction. Figures 2.1 and 2.2 provide us with examples of synchronic and asymmetric kind necessitation. Figure 2.1 depicts the fact that *scarlet*, among other kinds, kind-necessitates *red*. This is an example of a conceptual or metaphysical necessity, and our definition of 'kind necessitation' is clearly satisfied by this example. In any (conceptually or metaphysically) possible world whenever an object is scarlet it also is red. Plus, we have some intimacy here because an object's being scarlet explains how it is red (just as being 6′6″} explains how someone is tall). Similarly, Figure 2.2 depicts the fact that—and here I am stipulating—some maximally specific neuroscientific kind, say NS_1, kind-necessitates *pain*. Again, a person's being in NS_1 explains how or why it is in pain. These two examples appear to be very similar.

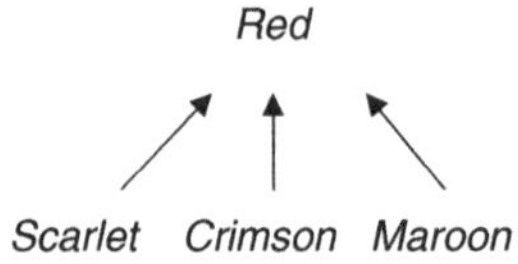

Figure 2.1 Specifications of *red*

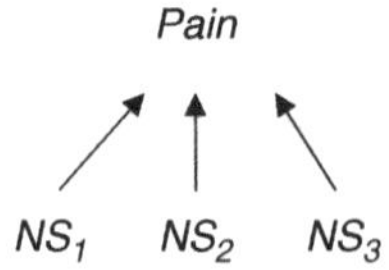

Figure 2.2 Maximally specific neuroscientific kinds

One difference between these two examples, however, is that *scarlet*, *crimson*, and *maroon* are not maximally specific color kinds, in the sense that there are more specific shades of *scarlet*, *crimson*, and *maroon*. But by supposition NS_1, NS_2, and NS_3 are maximally specific neuroscientific kinds, not allowing for greater specificity in neuroscientific terms. For rigor, I offer the following definition of a 'maximally specific *X*-kind':

> **Maximally specific *X*-kind:** An *X*-kind for which no finer specification can be made utilizing the predicates or classifications of science or conceptual scheme *X*, or any recognizable improvement (but not replacement) of conceptual scheme *X*.[2]

So, by supposition, the predicates of neuroscience cannot further specify NS_1, NS_2, or NS_3. Of course current science is not ideal, and it changes and (we hope) improves. The "recognizable improvement" clause of this definition is supposed to recognize this fact. A maximally specific *X*-kind is a kind that cannot be further specified by any future improvement of the conceptual scheme that is still recognized as the same conceptual scheme. This introduces vagueness and uncertainty—how much change can a conceptual scheme sustain while remaining the same?—but this is a common fact about issues concerning theory change and replacement. The clause also reflects the fact that kinds and their instances are not mere conceptualizations, but real-world phenomena. So whether or not a kind is maximally specific is not necessarily determined by our current predicates. In our example, I assume that neither NS_1, NS_2, nor NS_3 can be further specified by the current predicates of neuroscience or any recognizable future improvement thereof. We can remain neutral as to whether Figure 2.2 depicts an example of metaphysical necessity or whether a weaker modality, such as physical necessity, is involved here. But the idea is that in any metaphysically (or physically, etc.) possible world whenever an object is in NS_1 it also is in pain.

While these are paradigm examples of synchronic kind necessitation relations, there is an important difference between the two examples. This difference points to a very significant division between two varieties of synchronic kind necessitation relations more generally. In short, the difference is as follows: *scarlet* is a specific kind of *red*, but NS_1 is *not* a specific

[2] This definition also holds for substantival kinds. For example, species are the maximally specific biological kinds according to the traditional Linnaean taxonomy.

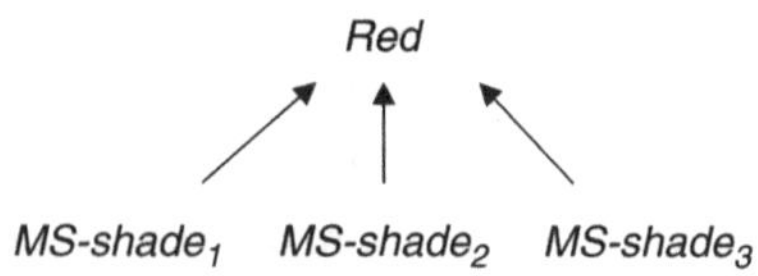

Figure 2.3 Maximal specifications of *red*

kind of *pain*.[3] The latter is bound to be a contentious claim, and the bulk of this chapter is devoted to defending it. But before defending it fully let me deflect a preliminary complication.

One might agree with this claim, but for the wrong reasons. One might think that *scarlet* is a specific kind of *red*, in part, because it is at the correct level of generality. In contrast, NS_1 is a maximally specific kind, but *pain* is a quite general kind. One might then think that NS_1 is not a kind of *pain* simply because it is too specific as far as psychological theorizing goes. But is this difference in specificity driving the judgment that NS_1 is not a kind of *pain*? No, it is not. And we can see this by substituting in maximally specific *color* kinds for *scarlet, crimson*, and *maroon*, as in Figure 2.3. Just as *scarlet, crimson*, and *maroon* kind-necessitate *red*, each of *MS-shade*$_1$, *MS-shade*$_2$, and *MS-shade*$_3$ kind-necessitates *red*. And each maximally specific shade of *scarlet, crimson*, and *maroon* is a (maximally specific) kind of *red*.

But why not say that NS_1 is a kind of *pain*, just like *MS-shade*$_1$ is a kind of *red*? I imagine that this will be a common reaction, so I will offer a cursory justification at the outset. My denial here is driven by the following assumption: two or more maximally specific neuroscientific kinds can, as far as we know, kind-necessitate the same maximally specific *pain* kind.[4] This, I take it, is what is meant by the widely accepted claim that *pain* is multiply realizable. This topic receives much more elaborate treatment in Chapters 4 and 5. For simplicity let us assume, implausibly, that *pain* is exclusively a phenomenological kind that varies only in intensity. (Nothing hinges on this assumption.) So a maximally specific *pain* kind is a kind corresponding to one maximally specific intensity value—e.g., *level 6.5 pain*. The assumption of MR tells us that more than one maximally

[3] If it will help, read 'variety' or 'type' for 'kind' here.

[4] Ehring (1996) and Walter (2007) make a similar point concerning the possibility of such a difference of realization without a difference in specification, which they also use to produce arguments similar to the one developed here (e.g., against Stephen Yablo's account of realization). I discuss Ehring's arguments further in §4.5.

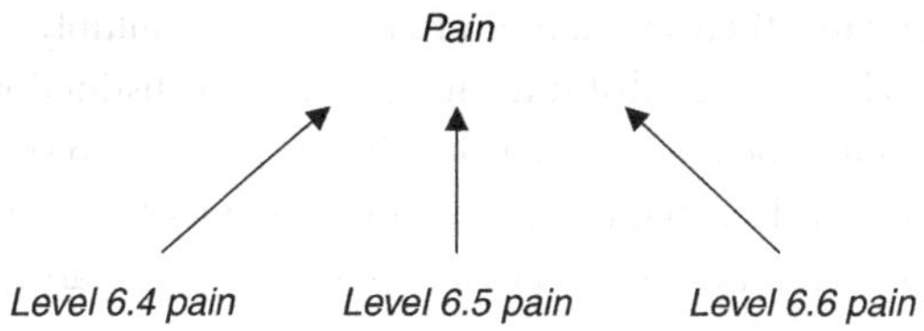

Figure 2.4 Maximal specifications of *pain*

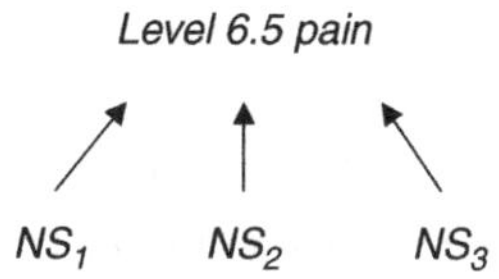

Figure 2.5 Multiple realizations of *pain*

specific neuroscientific kind can give rise to the same maximally specific *pain* kind, such as *level 6.5 pain*. But then these maximally specific neuroscientific kinds *cannot* be kinds of *pain*, as there can be sameness of maximally specific *pain* kind (i.e., *level 6.5 pain*) through difference in neuroscientific kind.

Again, we are assuming that *pain* is a purely phenomenological kind that varies only in its intensity. Figure 2.4 represents some of the (stipulated) maximally specific kinds of *pain*. Each of these maximally specific kinds is a distinct *pain* kind. Importantly, this is *not* necessarily true of the neuroscientific kinds represented in Figure 2.2. This point is made vivid by Figure 2.5, which contrasts nicely with Figure 2.4. Figure 2.5 represents the MR of *level 6.5 pain*. NS_1, NS_2, and NS_3 are not different kinds of *pain*. Rather, they are different neuroscientific kinds that give rise to the very same maximally specific *pain* kind.

Significantly, Figure 2.5 also informs us that *pain* itself is multiply realized by neuroscientific kinds. This is because multiple neuroscientific kinds (NS_1, NS_2, and NS_3) give rise to the same maximally specific *pain* kind (*level 6.5 pain*). *Pain* is not MR, however, simply in virtue of coming in varieties 6.4, 6.5, etc., any more than *color* is multiply realized in virtue of coming in distinct shades.

This characterization of MR is contentious, but it is defended in later chapters. Furthermore, the MR of *pain* has simply been assumed. If this assumption is false, then an example of a kind that is multiply realizable in this sense will have to be substituted in its place. The distinction that I am arguing for cannot be made (e.g., is incoherent) only if no such example exists. I hope it is clear that there are some examples that fit the distinction that Figures 2.1 and 2.2, as well as Figures 2.4 and 2.5, are supposed to capture.

The relevant difference between our examples involving *red* and *pain* should now be clearer. Each distinct kind of *scarlet* is necessarily a distinct kind of *red*. But each distinct neuroscientific kind is *not* necessarily a distinct *pain* kind, on our assumption about the phenomenological nature of *pain*. For this reason, we should introduce vocabulary to mark off these two varieties of synchronic kind necessitation. Let us say that *scarlet* specifies, or is a *specification* of, *red*. And NS_1 realizes, or is a *realization* of, *pain*. While not yet in a position to fully characterize specification and realization, we can make the following generalizations:

Kind specification: For any kinds *A* and *B*: If (and only if) *A* specifies *B*, then *A* is a kind of *B*.

Kind realization: For any kinds *A* and *B*: If *A* realizes *B*, then *A* is not a kind of *B*.

This claim has been put at the kind level, where I think it primarily applies. But we can also speak, in a derivative sense, of properties as realizing other properties.

Property realization: Property *x* realizes property *y* if, and only if, property *x* is of a kind that realizes a kind to which *y* belongs.

I will later explain why property *specification* is not an option.

I recognize that 'realization' has a well-established usage that some will think does not necessarily match the rather subtle distinction it is being employed to make here. As such, one of the primary goals of this chapter is to clarify, elaborate, and defend these claims and this usage. Chapters 4 and 5 offer a much more substantial defense of this way of conceptualizing realization. But, I think it is clear that no one should think that *red* is multiply realized simply in virtue of the fact that it comes in various shades, such as *scarlet, crimson*, and *maroon*.

There already is a significant literature on these two necessitation relations. The specification relation for adjectival kinds, I will argue, is the relation that holds among determinates and their determinables. And the realization relation has a large literature under that very name, particularly within the philosophy of mind. In recent years some have tried to assimilate the realization relation under the specification relation, but I will continue to argue that this is a mistake. The next few sections aim at explaining the specification relation for adjectival kinds and establishing that it is identical to the relation that holds among determinates and their determinables. The concept of specification or determination is then used as the key concept in articulating a property theory that partially explains the nature of properties, their kinds, and the relations among them.

2.3 Determination Dimensions and Property Spaces

Kinds related by the specification relation, such as *red* and *scarlet*, form a network. There is value in mapping these networks, because kinds belonging to the same network have important resemblances (as well as differences) and should be studied in a similar manner. But how can we determine, in general, whether two or more kinds are related by the specification relation? That is, how do we determine whether different kinds are also of the same kind? A convincing answer to this question is needed to motivate the current project. Fortunately, I believe that an answer is at hand. Kinds are related by the specification relation only if they are distinguished along the same *dimensions*. Let us call these the *specification dimensions* of those kinds. In the next section I argue that, for adjectival kinds, the specification relation is identical to that relation which holds among determinates and their determinables. So I prefer to call these the *determination dimensions* of adjectival kinds. These determination dimensions are meant to capture any feature of a kind according to which determinates and instances of that kind can differ from each other *qua* that kind. These can include causal, qualitative, dispositional, functional, or any other types of features that are part of the nature of that kind.

The terms 'determinable' and 'determinate' were coined by W. E. Johnson in Volume 1 of his classic work *Logic*. The terminology and corresponding relation did not garner a large literature, but prominent discussions and elaborations of the relation eventually could be found in Arthur Prior (1949), John Searle (1959), and Stephan Korner (1959). These discussions were primarily concerned with the logic of the relation itself and gave little attention to its application to actual philosophical problems. This changed starting in the 1980s when some metaphysicians—such as Cynthia and Graham Macdonald (1986), Stephen Yablo (1992), Douglas Ehring (1996), and Sydney Shoemaker (2001)—turned to the determinate–determinable (hereafter, *determination*) relation in the hopes of offering an account of mind–body relations and mental causation in particular.

Let us stick with our example involving colors in order to gain an intuitive understanding of determination dimensions. Colors, let us assume, are distinguished from one another only by their hue, brightness, and saturation. I recognize that these might not actually be the correct determination dimensions for *color*. If colors are distinguished by, say, wavelength, purity, and luminance instead, then these should be substituted in as the correct determination dimensions. What is important for present purposes is that there are some such dimensions, though we can disagree about what these are. If our assumption is correct, then *color* is a kind with three determination dimensions. Note that it is not just *color*, the quite general kind, that has these three determination dimensions, but also *red, blue*, and *yellow*, as well as *scarlet, sky blue*, and *canary yellow*. This is the central insight to understanding color specification: each *color* kind corresponds to some range of hue/brightness/saturation values with greater or less specificity. This insight can be generalized to aid in understanding specification for other determinable kinds.

Kinds sharing determination dimensions form important networks. *Color, red, blue, yellow, scarlet, sky blue, canary yellow*, etc. all belong to the same network in virtue of sharing the same determination dimensions. Let us introduce the term 'property space' to refer to such networks or subsets thereof.

> **Property space:** The metaphysical space that covers some range of physically compossible determination dimension values (along with the corresponding non-determinable necessities[5]) for a network or any partition of such a network.

Color is the ultimate kind for its network, as it is the kind that covers the entire range of compossible hue, brightness, and saturation values. *Red, blue, yellow*, and all other specifications of *color* cover only a limited range

[5] This component of kinds is discussed in the next section.

of *color's* determination dimension values. *Red, blue,* and *yellow* have their own property spaces, however, and these are proper subsets of *color's* property space.

Subsets of property spaces are themselves always property spaces. A property space point (i.e., location in a property space corresponding to some combination of maximally specific determination dimension values) is the limiting case of a property space for kinds that possess determination dimensions. In extreme cases, there are kinds that fall under no determinables and have no determinates themselves. Such kinds do not have determination dimensions, because they cannot vary nor are they maximally specific determinates of a determinable kind that varies. A fictional example of such a kind is *on*. A particle that is "on" interacts with other particles in predictable ways, such that *on* earns its keep as a legitimate kind, but there are no determinates of the kind *on* and no determinable that *on* falls under. Still, this kind is some abstraction, with an unanalyzable property space. Here the property space is exhausted by a single non-determinable necessity (to be explained shortly)—simply being on. I take it that some actual kinds (e.g., certain microphysical kinds) might be like *on* in this regard. Properties always belong to some kind(s), even if only to a non-determinable kind like *on*.

It is natural to describe kinds belonging to the same property space as being at the "same level of abstraction." In this sense, kinds can be at the same level of abstraction though differing in their specification. While *red* and *scarlet* are at the same level of abstraction, in that they share the same determination dimensions, they are not at the same level of specificity. *Scarlet* specifies *red*—i.e., its property space is a proper subset of *red's* property space. It is important to note, then, that kinds can differ in their degree of specificity without differing in their degree of abstraction. *Red* and *scarlet* both abstract away from everything but hue, brightness, and saturation, so for that reason they are at the same level of abstraction. It is surprising and significant, as we will see in Chapter 4, how common it is for a difference in level of abstraction to be confused with a difference in specificity.

We can also see why NS_1 does not specify *pain*, in our technical sense of 'specify.' We are assuming that pains vary only according to their intensity. In other words, intensity provides the only determination dimension for *pain*. But neuroscientific kinds are distinguished along altogether different dimensions, such as material type, activation state, and a whole host of other physical dimensions. With these distinct determination dimensions come distinct property spaces for neuroscientific kinds and *pain* kinds. Further, the fact that they possess different determination dimensions *explains* why there can be sameness of pain through neuroscientific differences. *Scarlet* and *red* provide a helpful contrast here: there cannot be sameness of red through difference in scarlet, because scarlets differ only according to their hue, brightness, or saturation, and these are all ways in which reds differ also.

Our examples of specification also show us that property spaces and the kinds associated with them can have either a *simple* or *complex* structure. A simple property space is one that has only one determination dimension. *Pain*, under the (obviously false) assumption that it is a phenomenological kind that varies only according to intensity, has a simple property space. A complex property space is one that has more than one determination dimension. *Color*, we have assumed, has a three-dimensional property space. So, we are assuming that *color* is a complex determinable.

It is important to recognize the difference between a kind that is complex *qua* determinable (i.e., has more than one determination dimension) and one that is complex in other senses. Not all kinds with complex structure are complex determinables. For example, the linear momentum of an object is the product of its velocity and mass. Still, velocity and mass are not determination dimensions of *momentum*. This is because objects with different velocities and masses can nevertheless have the same momentum (e.g., a 4 kg object moving 6 m/s has the same momentum as a 6 kg object moving 4 m/s). While *momentum* may be complex in the sense of having conceptual or metaphysical structure, it is not complex *qua* determinable. *Momentum* is determined along one dimension only—i.e., that dimension which can be measured in kg*m/s units.

The discussion thus far has largely been driven by intuitions. But I do not want to give the impression that determination dimensions are merely stipulated or are typically discovered as the result of *a priori* reflection. Rather, determination dimensions are supposed to pick out the ontological structure of properties and their kinds, and they typically exist independent of the classifying mind. At least this is true on our assumption of Property Realism. Even if Property Realism is not true, I assume that scientists believe that determination dimensions—which partially constitute the essences of kinds—are to be discovered by *a posteriori* investigation rather than by *a priori* reflection. As such, it is the business of science and other proven avenues of human inquiry to discover determination dimensions and the properties and kinds to which they give rise. The Nominalist can and should accept this methodological conclusion as well even if they deny that the discovered determination dimensions pick out the ontological structure, as opposed to, say, merely a highly useful way of conceptualizing an object. The point is that Property Realists and Nominalists can share the same scientific methodology and accept the same conclusions concerning the correct, productive, or preferable ways for conceptualizing and classifying the world. And the determination dimensions that a community discovers, in an ontologically neutral sense, partially fixes the way that community conceptualizes and classifies the world. The Property Realist and Nominalist differ only in the metaphysical conclusions that they draw from the determination dimensions they discover. The Property Realist judges that the world has an ontological structure corresponding to the determination dimensions accompanying their well-confirmed concepts or predicates, whereas the Nominalist does not read this structure into the fabric of the world.

To clarify matters let us discuss a particular example and further consider the differences between determination and realization. Let's again use the psychological kind *pain* as our example. Let us simplify as before and assume that *pain* is a purely phenomenological kind. For clarity we will call this kind *p-pain*, for *phenomenological pain*. As *p-pain* is a purely phenomenological kind, its determination dimensions will be the phenomenological dimensions that are distinctive of *pain*. Candidates for such determination dimensions could include intensity and duration. (Recall that in our earlier use of this example we assumed that *pain* had only one determination dimension—intensity.) Let's assume that these are the only determination dimensions for *p-pain*. Then the determinates of *p-pain* are the kinds corresponding to specific values (or ranges of values) of such intensity and duration. These are the classifications relevant to the science or study of *p-pain*. This is comparable to *scarlet* and *crimson* being determinates of *red* because they correspond to more specific values of hue, brightness, and saturation (the determination dimensions of *red*).

It is often claimed, in contrast to the phenomenological account just given, that many psychological kinds are functional kinds. Functional kinds are of particular interest to those engaged with cognitive science and other higher-level sciences more generally, because such kinds are widely thought to be multiply realizable and capable of being fruitfully investigated independent of realization-level study. So, let us now consider a psychological kind understood as a functional kind. We can continue with our example of *pain*, though this time assuming that it is a purely functional kind. For clarity we will call this kind *f-pain*, for *functional pain*. (I will leave it to others to settle whether *pain* is, as a matter of fact, identical to something like either *p-pain* or *f-pain*.) Matters are complicated because even if we limit ourselves to discussions within the philosophy of psychology, 'functional kind' has a few different senses. For present purposes I will understand a functional kind to be a kind that is characterized wholly relationally, specifically by its relations to causal inputs, other kinds within the same conceptual scheme (other folk-psychological kinds), and causal outputs. For example, and once again to oversimplify, let us suppose that *f-pain* is that kind of mental state which is caused by tissue damage, has certain distinctive relations to other mental states, and motivates behavior directed at avoiding the source of tissue damage and seeking treatment or healing. Just as the determination dimensions of *p-pain* are the phenomenological dimensions distinctive of *pain*, the determination dimensions of *f-pain* are the functional dimensions distinctive of *pain*. These would include the different types of tissue damage that can serve as causal input, the different types of relations to other mental states that are distinctive of *f-pain*, and the different ways of avoiding the source of the pain and seeking treatment or healing. For example, *f-pain* that is caused by burning, is accompanied by panic, and causes the subject to withdraw from the heat source is a specific kind of *f-pain*—a determinate of *f-pain*. This example should also suggest the form determination dimensions will take for alternative understandings of 'functional kind.'

Again, the concept of a determination dimension is quite general and applies to various kinds—kinds with essences that are phenomenological, functional, qualitative, etc. We can certainly disagree over the determination dimensions for a particular kind, but so long as this disagreement is reasonable, the very existence of such disagreement helps confirm that we share an intuitive understanding of the concept of determination dimensions and the task of discovering them. Determination dimensions are simply those essential dimensions of a kind along which instances of that kind can vary. Scientific kinds are of particular interest to metaphysicians, and I assume that the determination dimensions for such kinds typically are to be discovered by *a posteriori* investigation. In particular, the science of a given kind should provide us with the determination dimensions for that kind. However, for other kinds, and perhaps for some scientific kinds as well, determination dimensions might be discoverable by *a priori* reflection. Surely the determination dimensions of mathematical kinds are discoverable *a priori*, and perhaps the same is true of moral kinds (e.g., *duty*).

To see the wide application of this approach to taxonomization, an approach driven by the recognition of determination dimensions and the property spaces they generate, here is a list of some further kinds and plausible candidates for their determination dimensions:

- *sound*: pitch, timbre, and loudness
- *triangle*: three side lengths
- *mass*: grams (or other mass unit)
- *temperature*: degree (or other temperature unit)

The *sound* case is closely analogous to *color*, so there is no need to comment on it. It will prove helpful to justify the determination dimensions for *triangle*, however. Let triangle A and triangle B have the same three side lengths as measured according to some metric. At most one triangle can be formed from any three given side lengths, so we know that these triangles do not differ in their interior angle measurements or in any other triangle-relevant regard. All of the geometric and trigonometric facts about a triangle can be derived from the three side lengths. So with only the information that A and B are triangles with the same side lengths, we can appropriately conclude that A and B do not differ in their triangularity. Even if these triangles differ in mass, transparency, material constitution, color, or some other feature, they do not differ in their triangularity. Importantly, this shows that mass, transparency, material constitution, color, etc. are not dimensions along which *triangle* is determined. And while *triangle* is a three-dimensional (complex determinable) kind, *mass* and *temperature* are examples of kinds with simple property spaces.

Interesting questions can be raised regarding the status of complex property spaces. Not all of these are questions to be answered by the philosopher. Why must everything with a pitch also have a certain loudness? Why must everything with a hue also have a certain brightness? The philosopher is certainly not the appropriate person to answer these questions, and these are questions best left to the sciences. My attitude here is like that of many analytic philosophers—I humbly concede much ground to the scientists. This book presents a *logical* and *metaphysical* theory for thinking about the structure of kinds, and offers quite general guidance to the sciences and other proven disciplines. Specifically, the framework of property spaces nicely suggests the following roles for science:

1. Discover which properties and kinds exist.
2. Discover the determination dimensions of these kinds and the ontological structure of the properties which are their instances.
3. Discover and explain the bundling of complex property spaces (i.e., why certain determination dimensions must, or simply do, accompany one another).
4. Discover kind necessitation relations—including determination, realization, and causal relations—among specific kinds.

If Nominalism is correct, then these same enterprises can be pursued but without the ontological commitment to properties.

This section contains many assertions that are supposed to be intuitive but have yet to be fully defended. The remainder of this chapter presents a unified and formal theory of properties and their kinds. This theory accounts for the above assertions, and has other explanatory and theoretical benefits. The picture we have thus far is as follows. Properties are instances of various kinds. These kinds are represented and organized by their property spaces. And the structure of property spaces is provided by their determination dimensions. So in order to further understand the nature of properties and their kinds, we should pay more attention to the nature of determination. This is the task of the next two sections.

2.4 Determinates and Determinables

I have claimed that, for adjectival kinds, the specification relation is identical to the relation that holds among determinates and their determinables. In this section I will begin my argument for this claim, eventually providing a formal analysis of the determination relation in the next section. The account is of intrinsic interest, but also serves as the cornerstone of the property theory developed here.

In order to show that, for adjectival kinds, the specification relation is the same as the determination relation, I will first list ten features that are characteristic of the determination relation. This list overlaps significantly with our understanding of specification. Our analysis of the determination relation should be tested according to its ability to capture these features. Here are the features:[6]

1. *Kind level.* The determination relation holds between kinds.
2. *Canonical examples.* The following are some canonical examples of the determination relation: *red–color, scarlet–red*, and *circle–shape*. The first two examples combine to show that some kinds are determinates or determinables only relative to other kinds. *Red* is determinable relative to *scarlet*, but determinate relative to *color*.
3. *Specification.* For an object to be of a determinate kind is for that object to be, by metaphysical necessity, of the determinable kinds the determinate falls under *in a specific way*. For example, being scarlet is a specific way of being red, and being red is a specific way of being colored. This notion of specificity with regard to determinates–determinables can be compared and contrasted with other standard specificity relations—e.g., the disjunct–disjunction, conjunction–conjunct, and species–genus relations.
4. *Determinable instantiation requires determinate instantiation.* An object that falls under a determinable must also fall under some determinate of that determinable. Colored objects must be red or yellow or blue, etc. No object is merely colored *simpliciter*.
5. *Determinables follow determinates.* An object falling under a determinate also necessarily falls under every determinable that determinate falls under. Every scarlet object is also red and colored.
6. *Transitivity, asymmetry, and irreflexivity.* The determination relation is transitive, asymmetric, and irreflexive. Since *scarlet* is a determinate of *red*, and *red* is a determinate of *color, scarlet* is a determinate of *color*. But since *scarlet* is a determinate of *red, red* is not a determinate of *scarlet*. Furthermore, *scarlet*, like all kinds, is not a determinate of itself.
7. *Comparison under a determinable.* Determinates of the same determinable (as well as the objects instantiating them) admit of comparison in a way unavailable to pairs of kinds with no determinable in common. For example, an ordering or similarity relation obtains between determinate colors—*red* is more similar to *orange* than it is

[6] This list overlaps with many of the conditions provided by Johnson (1921: ch. XI), Ehring (1996: 470–471), and Armstrong (1997: 48–49).

to *green*. But no such comparison can be made between *red* and, say, *circle*.

8. *Super-determinates and super-determinables*. The transitive chain of a determinable and the determinates under it typically does not go on forever. At some point there is a kind that does not allow of further determination. Call such a kind 'super-determinate.' Super-determinate adjectival kinds are, to use our earlier terminology, maximally specific *X*-kinds. This is because, as I am arguing, for adjectival kinds the specification relation is identical to the determination relation. *Coca-Cola red* might be a super-determinate of *color*. Similarly, the chain of determinables that a determinate falls under typically comes to an end somewhere. Let us call a determinable kind that itself falls under no determinable 'super-determinable.' *Color* and *shape* might be super-determinables.
9. *No causal overdetermination by determinables and their determinates*. No effect is causally overdetermined simply in virtue of being caused by instances of a determinable and its determinates. For example, Mallory earned an 'A' for the course because her average was over 90%. However, this grade was not overdetermined by her average being exactly 93%. And *over 90% average* is determinable relative to *93% average*.
10. *Super-determinate exclusion*. The same object cannot fall under two or more super-determinate kinds of the same determinable at the same spatio-temporal location or, more generally so as to allow for non-spatio-temporal qualification, in the same precisely qualified way. For example, no object can be both some super-determinate shade of red and some super-determinate shade of yellow at the same place-time. Super-determinates exclude other super-determinates of the same determinable.

The similarities between specification and determination should be obvious with respect to most of these features. Specification in our technical sense is also a relation between kinds, it shares the canonical examples with determination, and, obviously, it is a kind of specification. Of all the features of determination, specification is most central to the relation. Just as the specification relation was limited to specification of a certain manner (i.e., along certain specification dimensions), determinates specify their determinables in the same limited manner. This limited sense of 'specification' is often overlooked in contemporary discussions of determination, in which realization is often seen as a species of determination or, conversely, determination is seen as a species of realization. I have already argued that this is a mistake, but in §4.4 I turn to specific confusions that have been made in the literature on mental causation and MR.

That said, the present understanding of determination in this more restrictive sense is not entirely original. In fact, I think that the present account is faithful to the original intent of Johnson and others. We find W. E. Johnson and Arthur Prior making similar points in their early writings on the relation. For example, Prior writes:

> Determinates under the same determinable have the common relational property, presupposing no other relation between the determinates themselves, of characterising whatever they do characterise in a certain *respect*. Redness, blueness, etc., all characterise objects, as we say, "in respect of their colour"; triangularity, squareness, etc., "in respect of their shape." And this is surely quite fundamental to the notion of being a determinate under a determinable.[7]

Prior's "in respect of" locution corresponds to our talk of levels of abstraction or determination dimensions. To say that redness characterizes objects in respect of their color is tantamount to saying that redness characterizes objects in terms of their hue, brightness, and saturation (if these are, indeed, the correct determination dimensions for *color*). In fact, I have chosen the example of *color* because we can trace it, and its determination dimensions, back to W. E. Johnson's original writings on the determination relation:

> Our familiar example of colour will explain the point: a colour may vary according to its hue, brightness, and saturation; so that the precise determination of a colour requires us to define three variables which are more or less independent of one another in their capacity for co-variation; but in one important sense they are not independent of one another, since they could not be manifested in separation.[8]

The determination dimensions are those manners in which determinates and instances of a kind can differ *qua* members of that kind—e.g., colors, *qua* colors, can differ only in their hue, brightness, or saturation. And I have strongly emphasized the role determination dimensions occupy in explaining the logical or metaphysical structure of kinds. However it would be a mistake to think that determination dimensions

[7] Prior (1949: 13). [8] Johnson (1921: 183).

exhaust the logical or metaphysical structure of kinds and their properties. Determinates and determinables often have structure in addition to these determination dimensions. Let us call these, the second type of structural features of kinds, their *non-determinable necessities*. Such features are so-called because they are features that each instance of a kind must have, but which do not allow for variation or specification within that kind. For example, triangles are three-sided, closed, plane figures. While triangles must have such features, these are not features along which *triangle* is determined. As Johnson properly noted, determinates under the same determinable differ in a particular way—i.e., along their determination dimensions. But determinates under the same determinable cannot differ with regard to their non-determinable necessities. Two triangles cannot differ in their three-sidedness as such a feature does not admit of degree or variation within the kind *triangle*. These non-determinable necessities, as components of kinds and their instances, are just as much abstractions as are determination dimensions. But for one qualification, determinates under the same determinable have exact similarity in their non-determinable necessities, though they differ with regard to their values under the same determination dimensions.

What is this one qualification? Sometimes a determination dimension for a super-determinable is fixed, providing a non-determinable necessity for some but not all determinates of that super-determinable. Take, for example, the super-determinable *shape*. *Square* and *triangle* are determinates of *shape*. One of the determination dimensions for *shape*, let's assume, is number of sides. Although *square* and *triangle* are both determinates of *shape*, there is no variance within each of those kinds with respect to number of sides. So though they are determinates of a common determinable, they possess different non-determinable necessities—three-sidedness is a non-determinable necessity for *triangle*, and four-sidedness is a non-determinable necessity for *square*. So, we must formulate our account of non-determinable necessities so as to account for these cases. Let us call non-determinable necessities that arise simply by holding some determination dimension value(s) of the presiding super-determinable fixed *derived non-determinable necessities*. Determinates under a common determinable can differ in their derived non-determinable necessities. Indeed, for super-determinate kinds what are typically determination dimensions turn, strictly speaking, into derived non-determinable necessities, since super-determinate kinds have a point-sized region in a property space that does not allow for variation. But other non-determinable necessities hold for all determinates of a super-determinable, and they are not derivative from its determination dimensions. Call these *absolute non-determinable necessities.* These are non-variable features that hold for a super-determinable kind as well as for all of its determinates. All beliefs, let us suppose, are regardings-as-true. Regarding-as-true, then, is an absolute non-determinable necessity of *belief*, in contrast with the dimensions like content and confidence which can vary across beliefs. We can then revise our generalization about non-determinable necessities: determinates under a common determinable have exact similarity in their absolute non-determinable necessities.

The theory presented thus far offers a two-component picture of the logical and metaphysical structure of kinds and their instances (properties). I have used the qualification "logical or metaphysical" in characterizing this structure, because I leave it as an open question what forms these determination dimensions and non-determinable necessities can take (e.g., functional, qualitative, etc.). The structure of kinds is provided by their determination dimensions and non-determinable necessities. This two-component structure is also found in the properties that are the instances of kinds. Properties possess specific values for each of their determination dimensions, as well as the non-determinable necessities of the kinds to which they belong. Further, there is nothing more to the structure of properties and their kinds but for these two components. *In this sense, then, properties and their kinds are abstractions. They abstract away from all detail except for that captured by their determination dimensions and non-determinable necessities.* Since properties and kinds are abstractions, they do not have things such as, say, non-determinable *contingencies* as part of their structure. Properties are abstractions from all contingencies, leaving only the determination dimension values and non-determinable necessities that are essential to the presiding super-determinable kind.

2.5 A Model and Analysis

We can systematize our observations about determination, and by sticking with simple examples we can construct mathematical models for our theory of determination. Each determinable that can be modeled as such is determined along n determination dimensions. We can construct

n-dimensional spaces by taking each dimension as an axis in our model. Instances of those determinables, the properties that are of those kinds, correspond to unique points in these n-dimensional spaces. (I am simply assuming that each property is super-determinate in this way. In §3.4 I consider the possibility that this assumption is false, and I show why this would not change the subsequent theory of kinds and properties.) The determination dimensions may be bounded on one end, bounded on neither end, or bounded at both ends. Also, these dimensions may be either continuous or discrete. Properties corresponding to the same point in a property space exactly resemble each other in all intrinsic aspects, since they have the same determination dimension values and cannot differ in their absolute non-determinable necessities. Again, recall that on the present theory these two components—determination dimensions and non-determinable necessities—exhaust the nature of properties.

Color kinds have been assumed to be three-dimensional. Hue, brightness, and saturation provide us with our x, y, and z coordinates for *color*. Again, the n-dimensional space that a kind spans is its property space. With such mathematical models we can easily explain the determination relation in terms of *subsets* of property spaces: the property spaces of determinates are proper subsets of the property spaces of their determinables. *Scarlet* determines *red* because *scarlet*'s property space is a proper subset of *red*'s three-dimensional property space. We can map these property spaces as shown in Figure 2.6.

Here the larger rectangular prism represents *red*'s property space, while the embedded rectangular prism represents *scarlet*'s property space. The property space of *color* itself is the most expansive range of compossible hue, brightness, and saturation values. Each point in this three-dimensional property space of *color* corresponds to a super-determinate *color* kind. And each instance of *color* corresponds to some point in this space (i.e., super-determinate *color* kind). Finally, color properties corresponding to the same point in this property space exactly resemble one another in all intrinsic aspects.

Points in property spaces represent super-determinate kinds. These are maximally specified kinds. But, I should emphasize, these are kinds nonetheless. Super-determinate kinds are unique, however, in that all of their instances are *exactly* resembling. Instances of determinable kinds, in contrast, can differ in their determination dimension values. Some might be tempted to think that because super-determinate kinds

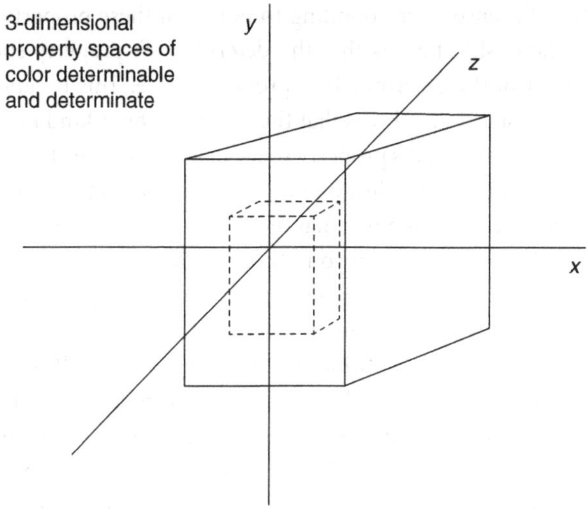

Figure 2.6. *Color* property space

are maximally specified, they are particulars. Particularity is often associated with maximal specificity, as the properties instantiated by particulars are always super-determinate (assuming that the world itself is not vague). But specificity should not be confused with particularity. Super-determinate kinds—having many possible instances—are just as much kinds, as opposed to particulars, as are the determinable kinds that they fall under.

With the *color* example in mind as a helpful heuristic, we can generalize necessary and sufficient conditions for determination as follows:

Necessary and Sufficient Conditions for Determination

Kind *B* determines kind *A* if and only if:

1. *A* and *B* have the same determination dimensions,
2. *A* and *B* have the same absolute non-determinable necessities, and
3. the range of determination dimension values for *B* is a proper subset of the range of determination dimension values for *A*.

These conditions should be acceptable to both the Nominalist and Property Realist, who need only differ over whether there are properties, in the Realist's sense, corresponding to points in these property spaces. The key relationship here is that the determinate's property space is a proper subset of the determinable's property space. This is captured by condition 3. Condition 3 shows that the *B* kind is the *A* kind *in a specific way*. There are points in *A*'s property space that are outside of *B*'s property space, so being of kind *B* is just one way of being of kind *A*. Conditions 1 and 2 are included to guarantee that the determinate's instances (properties) do not include features beyond those possessed by instances of the determinable, and vice versa. Properties and their kinds possess two types of components—determination dimension values and non-determinable necessities—so we need conditions to cover each of these. If *A*'s determination dimensions were not exhaustive of *B*'s determination dimensions, then it would be possible for properties corresponding to the same point in *B*'s property space not to exactly resemble in all intrinsic aspects. They could differ in the determination dimension value(s) not shared with *A*. Thus, condition 1 is required. And condition 2 guarantees that *B* possesses the absolute non-determinable necessities of *A*. For example, let us suppose that properties corresponding to points within the property space of *color* must, as a non-determinable necessity, also be spatially extended.

It might prove helpful to apply this same mathematical modeling procedure to another case. Take *rectangle* and one of its determinates, *square*. Rectangles differ from each other only in the lengths of paired opposing sides. Thus, *rectangle* is a two-dimensional determinable. Further, there is no privileged orientation for viewing rectangles, so that a rectangle that is 4×8 does not differ in its rectangularity from one that is 8×4. In order not to double-count types of rectangles, we can establish the convention of listing the larger of the lengths first (and in cases of ties, the order is obviously irrelevant). The property space of *rectangle*, with side lengths represented by the x and y axes and any derived non-determinable necessities shared by squares and rectangles (e.g., interior angles) omitted, is represented in Figure 2.7 as the area under the line. Squares are rectangles for which x equals y. So, *square*'s property space corresponds to the ray given by our equation $x = y$, where the domain of x is from 0 to infinity.

Mathematical models are tidy, but it is not obvious that all determinables can be so handily modeled. Consider, for example, the super-determinable *shape*. *Rectangle*, *triangle*, and *circle*, are all determinates of *shape*, but one might object that they cannot be represented by a fixed number of determination dimensions. The determination dimensions for shapes

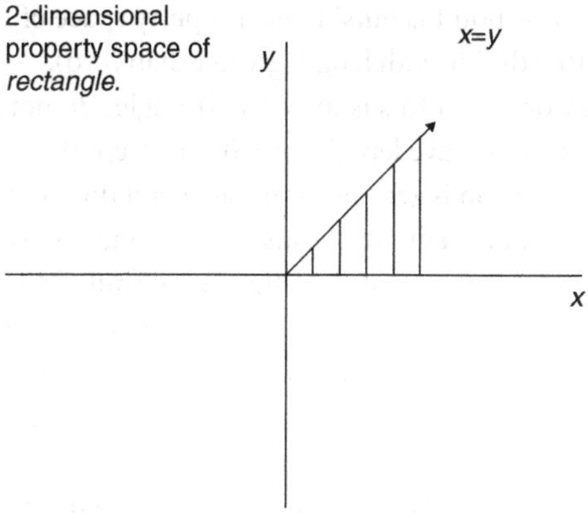

Figure 2.7. *Rectangle* property space

will need to capture features like the number of sides, lengths of sides, and their interior angles. For example, corresponding to each side for some shape there needs to be a distinct axis for its possible lengths. But rectangles, triangles, and circles have different numbers of sides, so it seems that they will require different numbers of axes to represent their possible side lengths. Since these determinates have different numbers of determination dimensions, they are not embedded in the same model and appear to have distinct property spaces. But they are all shapes!

Does *shape* then provide a counterexample to our two-component analysis of determination? Fortunately, our core analysis of determination applies equally well for the relationship between *shape* and its determinates. Instances of *shape* are abstractions, characterized by their determination dimensions and non-determinable necessities. Geometry, trigonometry, algebra, and calculus can provide us with the determination dimensions for *shape*. We easily recognize the number of sides and side lengths as two ways that shapes can vary. So let's allow an infinite number of determination dimensions corresponding to the number of possible sides a shape can possess, each with its own range of possible side length values. We then allow for zero (or no value) as a value along these dimensions, such that all triangles, say, have side lengths of zero for the determination dimensions corresponding to their fourth side, fifth side, and so on. This generates a common property space for all shapes. It also has the welcome consequence that *triangle*, though a determinate of *shape*, is not a determinate of *square*. For, all triangles have zero for their fourth side length, and this point is outside the property space of *square* (i.e., all squares have a fourth side with length greater than zero).

Some might worry that this is too easy. Triangles do not have a fourth side, which happens to have length zero. But I accept this. To say that the value for this dimension is zero *just is* to say that it does not have a fourth side. Or if it is not equivalent, we can just let "no value" be a possible determination dimension value. But then the worry shifts. Why do triangles have this fourth determination dimension, say, but not colors or any other arbitrary kind that intuitively belongs to a wholly distinct property space? If triangles have no value for this determination dimension, then why cannot colors as well?

I think this worry can be addressed if we approach things from the perspective of considering the super-determinables. Take the kind *shape*, a super-determinable considered in its full generality. Shapes can have various sides, with various lengths, and at various angles. We easily see that there are such varied, determinate kinds. So, we posit determination dimensions for *shape* to capture each such kind of variation. This generates a property space from the top down (i.e., by considering the super-determinable first, and then seeing the determinates it generates). *Triangle* naturally belongs in this space, with zero or no value for its fourth side, fifth side, etc. When we consider the super-determinable *color*, we see a parallel generality. They can vary along the hue, brightness, and saturation dimensions. There is nothing like side length along which they can vary. Note that some determination dimensions might not, by nomological or metaphysical necessity, allow for zero as a value. For example, there can be shapes without a fourth side, but maybe there cannot be colors without hues. But the legitimacy of these distinct bundles of determination dimensions should have strong intuitive appeal.

While we might not be able to construct a simple n-dimensional model of some property spaces, we still can provide an analysis of determination that covers kinds that are difficult, if not impossible, to model. The general idea is that super-determinables come with limited determination dimensions that span a range of values. Determinates are simply kinds that fall under the determination dimensions of some determinable, but have determination dimension values that span a proper subset of the determinable's determination dimension values. As with the kinds that are easier to model, 'being a proper subset of' is the critical relation and 'determination dimension' is the key concept.

2.6 Confirming the Analysis

A successful analysis of the determination relation should explain, or at least be consistent with, the ten pre-theoretical features of the relation detailed in §2.4. Other things being equal, the proposed analysis that does the best job of explaining, or is at least consistent with, these features is to be preferred. In this section I derive each of these features from the present analysis. For ease of explanation, the derivations that follow will use mathematically modeled determinables as examples. The key notion behind many of the derivations, 'proper subset of determination dimension values,' is shared by all determinates.

1. *Kind level.* It is simply built into the analysis that the determination relation holds between kinds.
2. *Canonical examples.* Our necessary and sufficient conditions do yield the verdict that our canonical examples—like *color/red, red/scarlet*, and *shape/circle*—are related as determinable to determinate. We can easily see this with our first pair. *Color* and *red* each has only three determination dimensions—hue, brightness, and saturation. Further, whatever non-determinable necessities are had by *color* are also had by *red* (though, it is unclear what these would be). And finally, *red*'s property space is a proper subset of *color*'s property space. *Color* spans all possible hue/brightness/saturation values, whereas *red* spans a limited range of hue/brightness/saturation values.
3. *Specification.* Our conditions for determination require that the property space of a determinate be a proper subset of the property space of its determinables. To be a proper subset the determinate must, with metaphysical necessity, specify a more limited range along at least one determination dimension. So, the determinate kind is a specific way of being the determinable kind. Further, the notion of a determination dimension explains the limited ways in which determinates specify the determinables they fall under. This contrasts with supervenience and realization relations, much discussed in Chapters 4 and 5, which are sometimes confused with determination.

   Our analysis also does not let so-called conjunctive kinds like *red-and-square* count as a determinate of *red*. *Red-and-square* does not have more precise values along the hue, brightness, or saturation dimensions than does *red*. Nor does it have more precise values along any of *shape*'s determination dimensions. This so-called conjunctive kind would correspond to subsets of distinct super-determinable property spaces as opposed to a unified property space. Nor does *scarlet-and-square*, say, determine *red*. For although *scarlet-and-square* has more specific color-relevant determination dimension values than does *red*, it also possesses shape-relevant determination dimension values. But our conditions for determination require sameness of determination dimensions, and *red* has no such shape-relevant determination dimensions.

Whether we should allow for so-called disjunctive kinds to be determined by their disjuncts is another matter. A so-called disjunctive kind would have disjunctive determination dimensions or would simply be a disjunction of entire property spaces. If there are disjunctive kinds, then one might think that *red*, say, is a determinate of *red-or-square*. I am deeply skeptical about such disjunctive kinds, however. Note that the disjunction here is a disjunction of real resemblances (i.e., determination dimensions or entire property spaces), not merely a disjunction of linguistic formulation. I am more tolerable of disjunctive kinds that belong to a common property space—e.g., *red-or-gray*—as the disjunct kinds share determination dimensions. My skepticism about more radical disjunctive kinds is based on the idea that kinds are supposed to mark real resemblances. And objects that satisfy a disjunctive predicate typically do not thereby share a real resemblance.[9] For example, a fire engine and a window might both satisfy the disjunctive predicate red-or-square, but they do not thereby share a real resemblance. But regardless of the status of disjunctive kinds, it is simply not true that *red* would belong to the same super-determinable property space as would *red-or-square*. *Red-or-square* would need to include additional determination dimensions besides hue, brightness, and saturation. Hence, given condition 1 of our necessary and sufficient conditions for determination, *red* cannot determine *red-or-square*.

[9] Armstrong (1978: ch. 14) presents a classic case for this claim.

4. *Determinable instantiation requires determinate instantiation.* No determinable can be instantiated without some determinate of that determinable being instantiated. This is because an instance of a determinable corresponds to a unique point in that determinable's property space. Because it is a determinable property space, there are proper subsets of this property space of which that point is also a member. The instance of any determinable is also an instance of every more determinate kind to which that point belongs. We know that there is at least one such determinate, because each point by itself represents a super-determinate kind. If we also admit kinds for every possible carving of a determinable property space, then the point inevitably belongs to other determinates as well. More plausibly, there are natural divisions of intermediate determinables to which this point belongs.
5. *Determinables follow determinates.* An instance of a determinate corresponds to a unique point in that determinate's property space. Because the property space of a determinate is a subset of each of its determinables' property spaces, each point in the determinate's property space is also located in each determinable's property space as well. So, necessarily, an instance of a determinate is an instance of every determinable that determinate falls under.
6. *Transitivity, asymmetry, and irreflexivity.* We can prove transitivity as follows. Assume that *C* determines *B*, and *B* determines *A*. Then the property space of *C* is a proper subset of the property space of *B*. Also, the property space of *B* is a proper subset of the property space of *A*. 'Being a proper subset of' is a transitive relation, so the property space of *C* is a proper subset of the property space of *A*. Also, they each share *A*'s absolute non-determinable necessities. Therefore *C* determines *A*, and the determination relation is transitive.

   Our analysis also entails that the determination relation is asymmetric and irreflexive. *B* determines *A* only if the property space of *B* is a *proper* subset of the property space of *A*. But if the property space of *B* is a proper subset of the property space of *A*, then the property space of *A* cannot also be a proper subset of the property space of *B*. So the determination relation is asymmetric. Also, the property space of *A* cannot be a proper subset of the property space of *A*, as nothing can be a proper subset of itself. So the determination relation is irreflexive.

7. *Comparison under a determinable.* Ordering and similarity judgments among determinates under the same determinable, and their instances, are also explicable on our framework. As a first pass, similarity between such determinates can be understood in terms of the nearness of their corresponding points in the property space. Our mathematical model explains why *orange* is more similar to *red* than it is to *blue*. Namely, *orange*'s property space points are closer to *red*'s property space points than they are to *blue*'s property space points.

These judgments of nearness are sometimes difficult, though, when more than one determination dimension is involved. We quickly discover that we cannot make judgments of similarity *simpliciter* for such determinables, but only similarity or ordering judgments relative to one determination dimension. This is particularly evident when one determination dimension makes more subtle distinctions (e.g., as measured in the pure number of such distinctions) than does another. For example, assume a world in which there are only two possible values for the brightness of colors: dim and bright. Let us pretend that in this world there is still as wide a spectrum of hues as there is in the actual world. Pick a point in this property space. This point has a neighbor that is one over on the hue scale (leaving brightness and saturation fixed) and another neighboring point one over on the brightness scale (leaving hue and saturation fixed). Is the color represented by the first point equally similar to each of these neighbors? It appears not. This is because the move along the brightness dimension is much greater than the move along the hue dimension. But when there are numerous possible values for each dimension, similarity judgments are often problematic. And there is no common scale according to which values along the different determination dimensions can be compared.

Similarity of determinates is best understood when limited to a single determination dimension. There is a common metric along which nearness of, say, hue can be evaluated. Minimally, when point B lies between points A and C according to hue, then the hue of A is more similar to the hue of B than it is to the hue of C. Simple determinables do not pose an incommensurability problem for their determinates as they possess only one determination dimension.

8. *Super-determinates and super-determinables.* Our model also clarifies the notions of a super-determinable and super-determinate. Super-determinables correspond to entire property spaces (i.e., the full gamut of compossible values along the determination dimensions). Values along these determination dimensions may be unbounded. In that sense at least, it is logically possible for a determinate to fall under an infinite number of determinables. For example, *4 kg mass* might fall under the determinables *under 5 kg mass, under 6 kg mass*, and so on ad infinitum. But there is still a super-determinable kind presiding over these property spaces—*mass*. Unrestricted property spaces are the super-determinables.

   Again, for ease of exposition and due to its intuitive appeal, I am assuming that every property corresponds to a unique point in a property space. Points in property spaces have precise values along each determination dimension. Therefore, every property is an instance of a super-determinate kind. While each property is an instance of a super-determinate kind, it is also an instance of every determinable that super-determinate falls under. In holding that every property has super-determinate values, this theory follows in the tradition of Locke and Berkeley. Though they may disagree about the status of general ideas, both agreed that everything in the world is super-determinate. There are no triangles lacking precise side lengths. Similarly, our properties, understood as abstractions, should not be confused with indeterminates. Every property has precise determination dimension values, though much other detail is abstracted away. This abstraction, understood in the Modern's sense as a substraction, is not to be confused with a lack of specificity. To repeat a common theme, *levels of abstraction are different than levels of specification*. This is the difference between invoking different determination dimensions or property spaces, on the one hand, and invoking different values within the same determination dimensions or property spaces, on the other.

9. *No causal overdetermination by determinables and their determinates.* An instance of a determinable kind never causally competes with the corresponding instance of its determinate kinds for the simple reason that an identity always holds for such properties. These individuation conditions are presented and defended in the next section.

10. *Super-determinate exclusion.* It is built into our individuation conditions for properties that the same object cannot instantiate different super-determinate kinds of a common property space at the same spatio-temporal location. These conditions are offered in the next section as well.

2.7 Individuating Properties

Our models and analysis provide us with great understanding of the determination relation that holds among adjectival kinds. But in addition to this application to kinds, we can also use this framework to give an account of properties. In particular, we can utilize our property space models to provide conditions for individuating properties and discovering their natures. I propose that three conditions are individually necessary and jointly sufficient for this task: the instantiating object, spatio-temporal location, and property space values (i.e., determination dimension values and non-determinable necessities). Remember that properties are assumed to be particulars, such as the color of this shirt here and now. This example successfully picks out a property, as each of these three conditions is met. The shirt is the instantiating object, here and now is the spatio-temporal location, and the specific hue, brightness, and saturation values (in addition to any non-determinable necessities) are the property space values.

The following identity conditions then suggest themselves:

Property Identity Conditions

Property p = property q if and only if:

1. p is instantiated by object x, q is instantiated by object y, and $x = y$;[10]

[10] Here I show my preference for a two-category ontology of objects and properties. I assume that properties are instantiated by objects, but some may favor free-floating properties or a single category trope ontology. See Schaffer (2003) for such an alternative. For those who favor such a one-category ontology, condition 1 can be dropped. This condition also has the effect of ruling out "trope piling," in which the very same object instantiates exactly resembling tropes at the same location. For sophisticated discussion of the possibility of trope piling, see Ehring (2011: ch. 3).

2. p is instantiated at st, q is instantiated at st', and $st = st'$;[11] and
3. p and q share precisely the same determination dimension values and possess the same absolute non-determinable necessities.[12]

Obviously, these conditions assume Property Realism. But even the Nominalist can accept the spirit of these conditions. Rather than identifying properties (in the Realist's sense), she may accept analogous conditions for the claim that, say, this object's being red is one and the same as its being scarlet. Scientists and others are often interested in such questions—e.g., whether a person's being in pain is one and the same as her being in a certain brain state—regardless of their position, or more likely their lack thereof, on the ontological status of properties.

Whether or not Property Realism is true, we should sometimes care about the individuation of instances of kinds. We should want to know when two different descriptions describe the same truth and when two different descriptions describe different truths. This is the specification and realization distinction all over again. 'This is red' and 'This is scarlet' are two descriptions of the same truth (when uttered in the same context), because *scarlet* specifies or determines *red*. 'She is in pain' and 'She is in NS_1' describe different truths, because NS_1 is not a determinate of *pain*. Whereas the above conditions assume Property Realism, the Nominalist can simply replace this talk of instantiation with some nominalistic-friendly expression like 'is true of.' But variants of these conditions are also relevant and fruitful for the Nominalist because she too should recognize that we could use different terms or concepts to express the same truths about an object, only with varying specificity of description.

The Property Identity Conditions make no explicit reference to kinds. However, we can also speak of objects instantiating kinds in virtue of their instantiation of properties. We then have the following kind-level version of our identity conditions:

[11] Non-spatio-temporal properties are individuated solely by instantiating object and qualitative nature (conditions 1 and 3). But to provide unified individuation conditions, let's allow having no spatio-temporal location to be an acceptable value for condition 2.

[12] Having no value for either determination dimensions or non-determinable necessities (but not both) is also a possibility. Our example of *on*, discussed in §2.3, presented us with a kind lacking determination dimensions. *On*-properties obviously have no determination dimension values, then, and only possess non-determinable necessities.

Kind-Level Formulation of Property Identity Conditions

> Object *x*'s instantiating kind *P* at spatio-temporal location *st* = object *y*'s instantiating kind *Q* at spatio-temporal location *st′* if and only if:
>
> 1. $x = y$,
> 2. $st = st'$, and
> 3. *x*'s *P*-instantiation and *y*'s *Q*-instantiation share the same determination dimension values and possess the same non-determinable necessities.

Significantly, condition 3 of the kind-level conditions can be met only if *P* and *Q* are the same kind, are related as determinate to determinable (or vice versa), or otherwise have overlapping property spaces. Only adjectival kinds standing in these relations can share an instance (i.e., property). This formulation is likewise amenable to the Nominalist. She would simply replace the instantiation talk with some more ontologically modest vocabulary, such as 'satisfying predicate P.'

Quite a bit has already been said about the role that determination dimensions and non-determinable necessities play in individuating properties. So some defense of the first two conditions should be offered. I take it as obvious that properties, conceived of as particulars, should be individuated by their nature and location. Conditions 1 and 2 combine to meet the location requirement. And I think that conditions 1 and 2 are both necessary. First, we cannot individuate properties by instantiating object and nature alone. This is obvious when we consider that super-determinates of the same determinable exclude one another, but one and the same object can, say, instantiate different super-determinates of *color*. For example, a tree can instantiate both some super-determinate shade of *green* and a super-determinate shade of *brown*. This is because these *color* properties are instantiated at different spatio-temporal locations of the tree. The location of a property is not determined by instantiating object alone.

One might think that condition 1 is unnecessary, however, as the location of a property can be given wholly by its spatio-temporal location. I resist eliminating condition 1, however, because I am not willing to rule out the possibility of distinct objects occupying the same spatio-temporal location. Also recall that objects without spatio-temporal location are considered to have the same spatio-temporal location for present purposes. And I understand super-determinate exclusion as the claim that no *object* can instantiate distinct super-determinates of a common determinable at the same spatio-temporal location. But it might be possible for distinct objects to wholly occupy the same spatio-temporal location (or occupy no spatio-temporal location at all) while instantiating distinct super-determinates of a common determinable at the very same spatio-temporal location (or, again, at no spatio-temporal location). So, condition 1 is also required.

Let us now apply these conditions to a specific example. Erin's shirt is some specific color "all over." It is true that her shirt is both red all over and scarlet all over, as scarlet is the specific way in which the shirt is red. According to our conditions, the property picked out by "Erin's shirt being red all over (for some temporal extent)" is identical to that picked out by "Erin's shirt being scarlet all over (for that same temporal extent)." This is because in both cases the same object, Erin's shirt, is of these kinds over the same spatio-temporal extent. The only thing that could then distinguish "them" is, speaking broadly, their qualitative nature. So, is there a qualitative difference here? Erin's shirt is one particular hue, brightness, and saturation combination (i.e., color), so it corresponds to only one point in the property space of *color*. Because this point is in both the property space of *red* and *scarlet*, this property is an instance of both *red* and *scarlet*. This red/scarlet property corresponds to one combination of determination dimension values and non-determinable necessities, so condition 3 is also met. This instance of *red*—the property—is then numerically identical to this instance of *scarlet*.

For contrast, consider an object that instantiates kinds that are not related in one of the three ways—identity, determination, or some other form of property space overlap—required to share an instance. Importantly, supervenience and realization relations amongst kinds are *not* sufficient to secure a property identity. Suppose that Erin instantiates some folk-psychological kind—say, she believes that Nebraska has a panhandle. She also instantiates some neuroscientific kind that kind-necessitates this folk-psychological kind.[13] Finally, assume that Erin

[13] In considering this example, let us put aside two worries. First, one could deny that the same object (Erin) instantiates both the folk-psychological and neuroscientific kinds. Second, one might object that the neuroscientific kind does not kind-necessitate the psychological kind due to wide content considerations. We could easily come up with another example of the right form for those who cannot put aside these worries.

instantiates both kinds over the same spatio-temporal region. A property identity holds only if the instantiations also correspond to the same point in a common property space. But this third condition is not met if the determination dimensions for the folk-psychological and neuroscientific kinds differ. For example, if the determination dimensions for the belief include dimensions like content and attitude, whereas the neuroscientific kind has radically different determination dimensions, like material type, activation state, etc., then they do not share a common property space. If the antecedent is indeed true, then these kinds cannot share an instance *even though* a kind necessitation relation holds between them. This would be an example of realization, not determination. Chapters 4 and 5 develop this point in great detail.

2.8 Determinates of a Determinable and Species of a Genus

The theory elaborated thus far applies to adjectival kinds and their instances (properties). But, with some modifications, it can be extended to apply to substantival kinds as well. The distinctive difference between adjectival kinds and substantival kinds is that instances of the former are abstractions whereas instances of the latter are *concreta* (even if they are non-spatio-temporal). For example, instances of *mass* are abstractions and instances of *tiger* are *concreta*. This difference in the ontological status of their instances requires a modification in the theory.

Both adjectival kinds and substantival kinds can be specified in limited ways utilizing the concepts and vocabulary distinctive to the conceptual schemes to which they belong. That is, there are specification dimensions for substantival kinds just as there are for adjectival kinds. I will henceforth use the term 'specification dimension' (and variants such as 'specifications,' etc.) for substantival kinds and reserve the term 'determination dimension' for adjectival kinds only. Specification dimensions and determination dimensions provide the variable, but essential, individuation conditions for substantival and adjectival kinds respectively. Consider, for example, the substantival kind *Homo sapiens*. *Homo sapiens* is a kind that belongs to a larger network of biological kinds. This network has its distinctive specification dimensions for classification, such as those concerning evolutionary relations. This network has developed into a very familiar and well-articulated taxonomy of substantival kinds.

Recall that on the two-component account of adjectival kinds and properties their natures are exhaustively provided by their determination dimensions and non-determinable necessities. For example, there is nothing more to a *color* property than its hue, brightness, and saturation values in conjunction with whatever non-determinable necessities belong to *color*. However, this is *not* true for substantival kinds. There is more to a human being, for example, than the specification dimension values and non-determinable necessities it possesses in virtue of falling under the kind *Homo sapiens*. For example, a human can be a philosopher, have brown hair, and weigh 190 pounds, though none of these features is relevant to the specification dimensions for the Linnaean System. So, unlike properties, the natures of instances of substantival kinds are not exhausted by their specification dimension values and non-determinable necessities.

A corollary of this is that objects (*concreta*) can fall under distinct, super-specifiable substantival kinds, whereas properties (abstractions) can fall under only one super-determinable. For example, instances of *color* do not belong to any other property space besides *color*'s property space (and its determinates' property spaces, which are subsets thereof). But human beings can belong to various substantival kinds that are not themselves related by specification or, more generally, do not share the same specification dimensions. For example, a human being can also belong to the kind *190 pound object* where this kind neither specifies nor is a specification of *Homo sapiens; 190 pound object* and *Homo sapiens* are kinds that belong to altogether different classification schemes.

This difference between adjectival kinds and substantival kinds was also recognized and emphasized by W. E. Johnson in his original writings on determination. Johnson made the point that since every adjectival kind falls under a unique super-determinable, adjectival kinds falling under distinct super-determinables are incommensurable. Johnson put the point as follows:

> Further, what have been assumed to be determinables—e.g., colour, pitch, etc.—are ultimately *different*, in the important sense that they cannot be subsumed under some one higher determinable, with the result that they are incomparable with one another; while it is the essential nature of determinates under any one determinable to be comparable with one another. The familiar phrase 'incomparable' is thus

synonymous with 'belonging to different determinables', and 'comparable' with 'belonging to the same determinable...[14]

The present theory differs only slightly from Johnson's position on this point. Johnson claimed that properties belonging to distinct super-determinables are incommensurable. I claim that comparisons are always along some determination dimension or non-determinable necessity. But perhaps distinct kinds can share at least one determination dimension while not belonging to the same super-determinable. If such cases are possible, then comparisons across super-determinable property spaces are possible. This qualification aside, Johnson's point stands.

I agree with Johnson that *color* properties and *sound* properties, say, are incommensurable, because they do not share any determination dimensions or (non-trivial) non-determinable necessities. But objects, unlike properties, can belong to substantival kinds that have no overlap in their specification dimensions or (non-trivial) non-determinable necessities. For this reason objects belonging to distinct, specifiable substantival kinds can nevertheless be commensurable. For example, I am a member of the substantival kind *Homo sapiens* and no rock is a member of a biological kind. Nevertheless, I can be compared to rocks in various ways, and some of the same predicates can be true of both of us—e.g., a rock and I could both weigh 190 pounds. So belonging to distinct, specifiable substantival kinds does not entail incommensurability. Objects falling under two distinct, specifiable substantival kinds can sometimes satisfy the same predicate and both be members of some third specifiable substantival kind.

[14] Johnson (1921: 175).

3

Objections and Responses

While I have gone some way toward providing a theory of the logical and metaphysical structure of kinds and properties, I acknowledge that some fundamental questions remain unresolved. Undoubtedly, there also are some points that need further clarification or are objectionable. In this chapter I respond to some of these objections and requests for clarification. Determination dimensions provide the backbone for this logical structure of kinds, so it would be wise to start with objections to this core feature.

3.1 Objections and Concerns about Determination Dimensions

3.1.1 Can Determinables Have a Dual Nature?

Someone might object that, contrary to the theory presented here, there are not unique determination dimensions and non-determinable necessities for every kind and property. The present theory claims that the nature of properties and their kinds is completely provided by their determination dimension values and non-determinable necessities. So on this theory a difference in determination dimension or non-determinable necessity entails kind distinctness. Likewise, a difference in determination dimension value or non-determinable necessity entails property distinctness. There is supposed to be a unique combination of determination dimensions and non-determinable necessities for every kind and property.

One might object here that the theory must be wrong because the very same kind or property can be accurately characterized by more than one set of determination dimensions or non-determinable necessities. Perhaps we can conceive of *color*, say, in terms of either (1) hue, brightness, and saturation, or (2) wavelengths and reflectance. Or perhaps *pain* can be conceived in either phenomenal or functional terms. These conceptions are supposed to generate distinct candidates for the determination dimensions or non-determinable necessities of the very same kind. Given the great emphasis I have put on these two components as providing and exhausting the nature of kinds and properties, this objection is potentially disastrous.

Fortunately, I think that an adequate rebuttal is at hand. A supposed counterexample to the uniqueness claim will purport to provide an example of a kind that fits two or more sets of determination dimensions or non-determinable necessities equally well. I believe that all such supposed counterexamples will fall under one of four categories. Each of these categories is consistent with the theory of kinds offered here, so there is no such counterexample after all. The four categories are as follows:

1. **Competition**: It is simply not the case that the two sets of proposed determination dimensions or non-determinable necessities are true of the kind in question. Rather, they are in competition and at most one of them accurately represents the nature of the kind in question and the other proposed set is to be discarded as not true to the nature of that kind.
2. **Supplementation**: It may be that both sets of proposed determination dimensions or non-determinable necessities are true, or partially true, of the kind in question, but only because neither exhausts the nature of that kind. In such a case the original set should be supplemented with determination dimensions or non-determinable necessities offered by the proposed counterexample.
3. **Conceptual or Linguistic Variation**: It might be that what appear to be distinct sets of determination dimensions or non-determinable necessities both exhaustively and accurately characterize the nature of a given kind. But, nevertheless, the supposed counterexample does not present a real alternative. Instead, the supposed counterexample might simply present the same determination dimensions or non-determinable necessities under the guise of different concepts or terms. This is possible because determination dimensions and non-determinable necessities, just like kinds themselves, can be picked out by distinct concepts and terms. This possibility allows for inter-theoretic and ontological reduction, as discussed in Chapter 5.

4. **Distinct Kinds**: The distinct sets of proposed determination dimensions and non-determinable necessities might both succeed in picking out kinds. However, they could be distinct kinds. This differs from Competition in that with Competition one of the sets simply fails to pick out any kind. For example, to use an earlier example, some might think that *pain* has purely phenomenological determination dimensions and others might think that it has purely functional determination dimensions. The reality might be that there are two distinct kinds here—*p-pain* and *f-pain*—and at most one of them is the referent of 'pain.' (Another possibility is that this presents an example of Supplementation, in which *pain* has both the phenomenological and functional determination dimensions.)

I believe that every supposed counterexample fits into one of these four categories.

Jessica Wilson (2009) argues against my claim that every kind has a unique set of determination dimensions. To the contrary, she thinks that scientific kinds typically have multiple determination dimensions that vary according to the science from which that kind is studied. To argue her case she uses my (2006) example, repeated here, of *color* and its supposed three determination dimensions—hue, brightness, and saturation. Wilson grants that these might be the determination dimensions for colors observed under normal light conditions, but she claims that consideration of alternate light conditions reveals that retinal spectral power distributions (SPDs) are necessary for individuating colors (and, hence, provide determination dimensions for *color*). She generalizes this point to conclude the following:

> In other words: determination dimensions may be science-relative. Note that there's nothing mysterious about the science relativity (more generally: context relativity) of determination dimensions, from a physicalist perspective. That different sciences may treat the same determinable as having different determination dimensions reflects that different sciences and their associated laws may treat the same phenomena at different levels of metaphysical grain.[1]

Remember, though, that determination dimensions, in conjunction with non-determinable necessities, are supposed to provide the essence of a kind. Wilson's claim has the unsettling consequence, then, that the

[1] Wilson (2009: 163).

essence of a kind—what it is to be of that kind—varies according to the science from which that kind is studied. Of course different sciences can, in one sense, treat "the same phenomena at different levels of metaphysical grain." For example, different sciences—such as neuroscience and cognitive psychology—can investigate human behavior (the same phenomenon) at different levels. But it is an altogether different claim to think that each science is positing the same *kinds* to explain human behavior and that the perspectives of these different sciences provide *different* determination dimensions, and thereby essences, for the *very same* kinds. It is the latter claim that I deny if these are truly autonomous sciences at different levels of abstraction. (More on this in Chapters 4 and 5.)

I do not want to defend any substantive claim about the actual determination dimensions for particular scientific kinds. All my candidate determination dimensions are simply plausible dimensions offered for illustrative purposes. To clarify, I am not concerned with defending any substantive thesis about the nature of *color, pain*, etc. So it very well might be the case that *color* possesses determination dimensions different from the three—hue, brightness, and saturation—that I have assumed (not argued for) in illustrating the concept of a determination dimension. It might be that the determination dimensions for *color* are SPDs, instead.

I see this supposed counterexample, and any like it, as falling under one of the four categories just presented. First, it might be that there is one kind, *color*, and SPDs compete with hue, brightness, and saturation (among other candidates) for providing the determination dimensions of that kind. *Color* possesses one set of determination dimensions or the other. Second, there could again be this one kind, *color*, and it possesses determination dimensions of both types (supplementation). Third, the supposedly competing candidates for determination dimensions might actually be two different ways of describing or referring to the same determination dimensions. And fourth, it might be that there really are what are called *color* kinds with determination dimensions of these diverse types, but the kinds themselves are distinct. In effect, when theorists from these different scientific perspectives talk about color, as they understand it with their different determination dimensions, they are equivocating on the word 'color.' And this fourth option is often the case whenever, as Wilson puts it, different sciences study the same phenomenon. However, the different disciplines frequently have their own vocabulary for the kinds they study, so there is not an issue of equivocation. The *concreta* they are studying are the same, but the abstractions those sciences posit—their kinds and properties—are typically distinct. The determination dimensions and non-determinable necessities that they successfully posit reveal that these sciences are examining kinds with distinct essences. Therefore, the kinds are distinct.

The four categories used to assimilate cases of supposedly dual-nature kinds provide us with a practical, scientific way of legitimizing the claim that kinds have unique determination dimensions. But more abstract, metaphysical considerations also support this point. Kinds and their instances just *are* abstractions. Determination dimensions and non-determinable necessities provide the different abstractions. A difference in determination dimension or non-determinable necessity is a difference in abstraction; a difference in abstraction is a difference in kind; therefore kinds have unique determination dimensions and non-determinable necessities. These abstractions are mind-independent, and we must always be careful that we do not fall into the mistake of taking our conceptualizations for the abstractions themselves. This is why our conceptualizations—e.g., the determination dimensions we posit—can compete with each other, supplement each other, be notational variants of one another, or coexist as marking distinct kinds. Mind-independent facts about the world largely determine which candidate should prevail.

I have provided two reasons for thinking that kinds cannot have a dual nature. First, cases of supposed dual nature are to be assimilated into one of four categories (competition, supplementation, conceptual or linguistic variation, or distinctness). Second, and more fundamental, kinds (and properties) are abstractions that simply abstract away all features but for one nature. One might accept these points about kinds, but still insist that a property can have a dual nature. Let's examine one such proposal.

David Robb (1997) argues that mental tropes are identical to physical tropes, even if mental kinds are not identical to physical kinds. He proposed this as a solution to causal exclusion worries. He accepts that all physical effects have only physical causes—in particular, only physical properties are efficacious—but he also wants to vindicate the causal efficacy of mental properties. His proposal is simply to identify mental tropes with physical tropes. I find this proposal unconvincing, among other reasons, because there is no independent reason to think that the tropes are identical, nor is a general theory of trope individuation offered. He does not make a specific case as to why, apart from avoiding exclusion worries, such an identity is legitimate. As such, I see the threat of epiphenomenalism as still looming for him. But I also think that there are specific reasons against such an identification, so let me say a bit about why I think that this trope identity proposal is unacceptable.

In effect, I see Robb as not taking seriously enough the idea that tropes are abstractions. I have offered a theory of properties (tropes) according to which this notion of abstraction is cashed out in terms of the determination dimensions and non-determinable necessities for the kinds to which they belong. Since these differ for physical kinds and mental kinds—on the non-reductive view that Robb and I are both assuming—then there is a non-identity for the tropes as well.

The best sense that I can make of the trope monist view is that these mental/physical properties are abstractions that include both mental and physical "content," almost to the point that they are really just bits of *concreta*. By "content" I here mean whatever is left when everything else is abstracted away. On the theory I have offered, the content includes only determination dimension values and non-determinable necessities, but others can have their own views as to what this includes. We can extract some of Robb's commitments from passages like this:

> Now if trope monism is true, a given mental type is a set of physical tropes. But multiple realizability entails that these physical tropes do not themselves resemble one another in the way that members of a *physical* type must: they will be wildly dissimilar physically. So the mental type is not itself a physical type (hence *Distinctness*), though of course it has many physical types as subsets: these are just the physical types that 'realize' the mental one.[2]

This passage suggests that the trope is simultaneously a mental and physical abstraction. On this view mental tropes of the same super-determinate mental kind exactly resemble each other *qua* mental, but they also contain physical content that can differ.

But the whole idea of tropes is that they are abstractions that can ignore this supervenience detail. Not only *can* cognitive psychology concern itself with properties (tropes) that abstract away from these physical differences, cognitive psychology *does* concern itself with such properties. And conversely, even though the mental content supervenes (let us suppose) on the physical content of the trope, physicists abstract away from

[2] Robb (1997: 188).

this mental content. As such, this content is not part of the trope at all. We see this in that the generalizations of cognitive psychology and physics do not concern themselves with the content (e.g., determination dimensions) of the other. For example, the (present) laws of physics do not refer to nor do their kinds vary along dimensions like phenomenal consciousness. Physics abstracts away from that supervenient detail, concerning itself with tropes lacking that content.

Here is another way to reach the non-identity conclusion: As Robb claims, his mental tropes of the same super-determinate mental kind do not exactly resemble *qua* physical. But, they do exactly resemble *qua* mental. But resemblance and properties (tropes) go hand in hand. If there is this real (exact!) psychological resemblance among the various mental tropes of the same super-determinate mental kind, then there are such tropes that exactly resemble *simpliciter*. But Robb's mental tropes of the same super-determinate kind do not exactly resemble *simpliciter*; they exactly resemble *qua* mental but differ *qua* physical.

As far as these objections are concerned, there is nothing distinctive about Robb's monistic proposal. These same objections apply to any supposed case of a dual-nature trope. Tropes are abstractions; the abstraction provides us with the nature. If there are truly two natures, then there are two abstractions. And if there are two abstractions, then there are two tropes. The best sense that I can make of Robb's proposal is that his tropes possess a composite nature, rather than a dual nature, in which the same trope is supposed to have both aspects at the same time (with neither the mental nor physical content providing an exhaustive nature). But this is to see the tropes as more concrete than they in fact are. In fact, the psychological and physical tropes abstract away the content of each other. Robb's theory fails to group together—as instances of the same super-determinate kind that exactly resemble each other—all the tropes that exactly resemble *qua* psychological, on the basis that they possess physical differences.

Realism about determination dimensions and properties with such structure leads to more serious metaphysical concerns over Property Realism. I admit that I have avoided directly taking on that metaphysical task. By and large I have simply assumed that properties are abstract particulars. I do think, however, that the trope theory is also a consequence of the property theory advanced here. I have just explained how on this theory properties and their kinds are abstract, with the determination dimensions and non-determinable necessities providing the level of abstraction. Properties are also particular, rather than universal, on this theory because a kind like *red* can share an instance—numerical identity—with *scarlet* on one occasion and *crimson* on another. A *scarlet* universal would be distinct from the *crimson* universal, however, thwarting such a possibility.

I do not think that defending Property Realism is essential to the task that I have set for myself here. Whether or not we are Property Realists, we should all acknowledge that there are statements akin to property ascriptions that are either true or false—e.g., "This banana is yellow" and "This banana is colored." Sometimes, in our more theoretical or inquisitive moments, we want to know whether two such statements are related in very interesting ways—e.g., whether the truth of one statement entails the truth of the other; whether both statements make the same claim about the world, only in different terms; etc. So I have advanced a theory that articulates and explains the conditions under which we have a genuine difference in kind or property, when we have genuine sameness of kind or property, and when we have a mere difference in degree (specificity). But I defer to others to prove whether Property Realism is true.

3.1.2 *What is the Epistemology for Discovering Determination Dimensions?*

Given the defining role that determination dimensions and non-determinable necessities play in this theory, and given my commitment to there being a unique set for each kind, it is natural to wonder how we do and should discover the determination dimensions and non-determinable necessities for specific kinds. Unfortunately, I believe that there is not a good answer to this general question. I do not think that there is any unified epistemology for discovering the determination dimensions and non-determinable necessities for the kinds or properties of the various sciences, mathematics, ethics, and all other conceptual schemes. But however disappointing this answer might be, it is more than reasonable. Since determination dimensions and non-determinable necessities provide the essences of their kinds or properties, I am simply claiming that there is no unified procedure for discovering the essences of kinds or properties across all conceptual schemes. And surely this is correct.

Philosophers often oversimplify matters here, I think, by erroneously believing that all kinds, or all kinds within a large domain, are of a similar (generic) nature. I will give two such examples. Some have claimed that the nature of all contingent kinds is exhausted by their causal powers.[3] And within the philosophy of mind it was recently, if not still, widely thought that psychological kinds are functional kinds. Many were indiscriminate in making such claims, however, and it is unlikely that for any conceptual scheme as wide and diverse as psychology *all* of its kinds (even putting aside qualia) would have exclusively functional essences. So, I advise more caution here. We should recognize that the nature of kinds can be quite diverse, and we should be reluctant to generalize that all kinds, even within some conceptual scheme, are of a certain (generic) nature.

I am willing, however, to offer some very general advice concerning the epistemology of kinds. Kinds belong to conceptual schemes which typically (if not always) include laws, generalizations, or explanations formulated in its kind terms. Some have recognized this connection between kinds and the explanatory work of the conceptual schemes to which they belong, arguing that nothing is a kind unless its corresponding kind term essentially occupies such a role in the laws or other explanatory work of its conceptual scheme.[4] And a simple methodological principle follows from this: to discover kinds, look for the kind terms that are necessary for formulating successful laws, generalizations, or explanations. But not only do these explanations provide us with a guide to discovering kinds, they also can guide us toward the essences of those kinds. For example, if generalizations or laws about *pain* take its intensity as a variable, then this is some reason to think that intensity is a determination dimension of *pain*. Or, to take an example from Dretske (1988), the fact that there are generalizations about *sound* that critically depend on its pitch also suggests pursuing pitch as a determination dimension of *sound*. These sciences are not infallible however—not even at articulating the determination dimensions for false theories!—so we should remember that determination dimensions can be provided by a recognizable improvement of a current conceptual scheme.

3.1.3 Are Simple Property Spaces Informative?

Determinable kinds can be either simple or complex. Some might object that while the present theory is informative when it comes

[3] Shoemaker (2003: ch. 10).

[4] Though he is concerned with natural kinds in particular, Fodor (1974) is representative here.

to complex determinables—with their multidimensional property spaces—it does not shed light on the nature of simple determinables. The theory has it that determinates simply have property spaces that are proper subsets of their determinables' property spaces. But if the determinable is simple, then it has only one determination dimension and its determinates are simply more specific kinds along that dimension. Then, isn't this theory merely saying that simple determinates are more specific versions (or determinates!) of their determinables? The "theory" of determination then seems completely uninformative for simple determinables.

There is some truth to this objection, in that I offer no account of resemblance along a determination dimension. I do not see this as at all problematic, however, as resemblance-in-a-respect, which is at the core of sameness of kind, *is* ultimately unanalyzable. One could go down a metaphysical path, to be sure, and claim that resemblance-in-a-respect is to be explained, say, by instantiating the same universal. But the buck must still stop somewhere—the distinctness of the universals would be just as primitive as, on my view, the distinctness of resemblances-in-a-respect. Such extravagent metaphysical posits would not help *explain* sameness of kind, as the question "In virtue of what (if anything) do these things instantiate the same universal?" remains for them just as the question "In virtue of what (if anything) do these things resemble-in-a-respect?" remains for me.

But our account of determination, even for simple determinables, still is informative. We can make this case both positively and negatively. On the positive side, our theory has it that kinds and their instances are abstractions with determination dimensions approximated by the theories and semantics for their kind terms. Sometimes kind terms are explicitly theoretical such that their determination dimensions are also explicit. Determination dimensions, like the kinds they give rise to, are mind-independent, and successful reference requires only an approximate fit between theory and reality. Further, the theorized determination dimensions can alter over time yet still maintain reference to the same kinds. How much alteration they can sustain while still maintaining reference is a vague matter—it is the difference between theory change and theory replacement. Other times kind terms are not yet theory-laden, but they nevertheless pick out determination dimensions for familiar, Kripke–Putnam causal theory of reference reasons.

For example, just like 'water' might pick out something that is essentially H_2O, 'color' might pick out something that essentially has a hue, brightness, and saturation. The positive claim is that determination is a relation between kinds at the same level of abstraction. On the negative side, this theory says what determination is not. In particular it is to be contrasted with realization, a necessitation relation between kinds at different levels of abstraction. So, for example, the multiple realizations of a simple determinable should not be mistaken for determinates of that kind. In §4.4.1 this view is defended against those who deny this. This very disagreement shows that this account of simple determinables is informative.

3.1.4 Do Tropes Have Determination Dimensions?

According to the property space model, properties correspond to points in a property space, so they have determination dimension values. This is built into the property individuation conditions from §2.7. But some might object that determination dimensions apply only to kinds, not to properties. Determination dimensions are dimensions along which a kind can vary, one could point out, but tropes cannot vary. So tropes do not possess determination dimensions.

There is some truth, but also some confusion, in this charge. The truth is that tropes cannot vary and in that sense do not have determination dimensions. But, as specified in §2.7, they do have determination dimension *values*. A *color* property, for example, does not vary in its hue, brightness, or saturation. But it does have a precise hue, brightness, and saturation value. So when speaking of properties I talk of determination dimension *values*; when speaking of kinds I talk of determination dimensions or ranges of determination dimension values.

Determination dimensions not only contribute to the structure of kinds, they also contribute to the structure of properties. Properties are taken to be abstract particulars, as defined in §1.2. The *abstraction* of a property is determined by the kinds—the determination dimensions and non-determinable necessities—to which it belongs. A *color* property is nothing but, say, a particular hue-brightness-saturation value. All other details, but for the values for determination dimensions for *color*, are subtracted away. In this sense, determination dimensions structure properties as well as kinds.

3.2 Are Properties or Kinds More Fundamental?

We have been presented with a theory of both properties and their kinds. Properties are instances of kinds, where this is understood as correspondence to a point in a kind's property space. Properties and kinds go hand in hand in this way, even though they belong to different conceptual or ontological categories. Though they have been presented as a package deal, we might wonder which is more fundamental—properties or kinds? Priority could be given to one over the other in at least the following three ways:

1. **Ontological Priority**: Are kinds ontologically dependent on properties or are properties ontologically dependent on kinds? Or does no relation of ontological dependence hold between properties and their kinds?
2. **Explanatory Priority**: Does an object's being of a certain kind explain its having a certain property, or does an object's having a certain property explain its being of a certain kind? Or does no relation of explanatory priority hold between properties and their kinds?
3. **Methodological Priority**: Should investigation (scientific, conceptual, etc.) begin with kinds or with properties? Or is neither to be given methodological preference?

Let's address each of these questions concerning priority.

ONTOLOGICAL PRIORITY

I have simply assumed Property Realism—that properties are irreducible parts of the world that are ontologically distinct from concrete particulars and sets thereof. Though I have not given any argument for this, the reasons that could be offered are, I think, familiar. Properties are often part of the spatio-temporal world and they make a causal difference. Properties are also things that we can perceive or measure with instruments. It seems that we can gesture towards them (if not literally point to them), linguistically refer to them, and name them. They have the markings of the ontologically real.

But what about kinds? Are kinds irreducible inhabitants of the world or do they otherwise exist independent of the classifying mind? If so, then Kind Realism is true. If not, then the ontological priority question is quickly answered in favor of properties, as kinds have no ontological standing whatsoever. So, what is the ontological status of kinds? Thus far I have been rather loose in talking about kinds, and it is now time to tighten up our understanding of them.

I have claimed that a kind is picked out by a region of a property space. As I understand them, property spaces are theoretical constructs, constructed by hypothesizing or discovering the determination dimensions of actual, or physically possible, properties. Property spaces can then be thought of as akin to predicates that succeed in picking out actual, or physically possible, properties. Significantly, there could be theoretical constructs just like property spaces but for failing to pick out actual, or physically possible, properties. These are akin to predicates that do not pick out properties—what we can call *mere predicate spaces*. Property spaces and mere predicate spaces have the same ontological standing, however. Both are theoretical constructs. Property spaces earn the title 'property' in virtue of representing actual or physically possible properties, whereas mere predicate spaces do not represent any actual or physically possible properties. This usage coheres with an interest in properties in a practical sense—as part of the fabric of the actual world.[5] For example, there is a property space for *color* because there actually are *color* properties. But there is not a property space for *phlogiston*, as there are no *phlogiston* properties nor are they physically possible (i.e., they cannot be instantiated consistent with our laws of nature). There was a phlogiston theory, however, and *phlogiston* was a hypothesized kind. So, there are predicate spaces corresponding to *phlogiston*. But since there are no actual or physically possible properties corresponding to this predicate space, it is a mere predicate space.

We can also speak of 'predicate spaces' in a different sense. Sometimes a predicate applies to properties from distinct property spaces with diverse determination dimensions. Such a predicate—e.g., 'beautiful'—picks out properties, but not properties that belong to a single property space. We can also speak of such predicates as generating a predicate space, though this is in a different sense than discussed in the previous paragraph.

I have spoken of kinds as being "picked out by" or "represented by" property spaces. But how should we think of this relationship between

[5] This restricted use of 'property space' is somewhat arbitrary. One could choose to use 'property space' in a more liberal sense, granting the label to any space that picks out *metaphysically* possible properties.

property spaces, understood as theoretical constructs, and their corresponding kinds? Here are two possible answers:

Kind Realism: Property spaces are merely theoretical constructs, but kinds are ontologically real and exist independent of the classifying mind. Property spaces are ways of *representing* kinds. A kind is the *class* of all the possible properties corresponding to a given property space, or the *universal* corresponding to a property space, etc.

Kind Nominalism: Kinds *just are* property spaces. So, kinds are simply theoretical constructs that pick out actual or physically possible properties. Some kinds might nevertheless be more *natural* than others—e.g., by picking out groupings that figure in law-like generalizations. Kind Nominalists can certainly differ over their standards for hypothesizing or discovering the determination dimensions of kinds, and they might favor pragmatic or parochial considerations in dividing properties into kinds.

The account of determination and property individuation offered here is compatible with either Kind Realism or Kind Nominalism. I prefer to simply identify kinds with property spaces. So, I am a Kind Nominalist and will assume this position in what follows. One challenge for the Kind Realist is to provide work that both needs to be done and can only be done by kinds, as understood in the Realist's sense. I happen to think that we can get by with only properties. Further, accepting Kind Realism commits one to ontological oddities like classes, universals, or *possibilia*. On the other hand if kinds are merely theoretical constructs, then the question of ontological dependence between kinds and properties simply does not arise. I leave it to the Kind Realist to answer the question of ontological priority on the assumption that kinds are ontologically real.

EXPLANATORY PRIORITY

So long as Property Realism is true but Kind Realism is false, it also seems that belonging to the same kind is to be explained by property similarity (rather than property similarity being explained by sameness of kind). But even for those neutral with respect to the question of Kind Realism, it should seem obvious that properties explain kind membership and not vice versa. We should understand that some object is of the kind *red* because that object instantiates a *crimson* property. Moreover, we are to understand that the object is of the kind *crimson* by recognizing that the object instantiates a *crimson* property. Though some Kind Realists will certainly disagree, I simply assert that it is an inversion of the explanatory order to claim that an object is crimson because it is of the *crimson* kind.

METHODOLOGICAL PRIORITY

At issue here is whether investigation should proceed by first searching after properties in order to discover or construct kinds, or whether instead we should begin by hypothesizing kinds to discover the properties. Pursuing the first route means that we first directly turn to the world and its properties and then, after that examination, we hypothesize determination dimensions (and thereby kinds) in light of our observations. The latter route is the reverse: we hypothesize determination dimensions and then search the world to see if there are any properties that fit them. As with all methodological questions, we should answer with whatever works. In fact, each approach is effective and, I think, they are necessarily interconnected. So, neither approach should be given methodological priority.

Scientific investigation is always a combination of data collection and theorizing. This process is typically iterated until a reflective equilibrium is reached. Direct investigation of properties, typically by empirical observation aided by instruments, is an example of data collection, whereas hypothesizing determination dimensions is a matter of theorizing. It is obvious that, to some extent, the study of scientific kinds is inevitably intermingled with the investigation of their instances (properties). This is especially true if one thinks that the essence of a scientific kind is determined by the role it plays in laws and, more generally, in its explanatory role within a theory or conceptual scheme.

That said, it is evident that in some cases properties were first discovered or known, and their kinds and determination dimensions discovered or constructed after the fact. This is the case with folk kinds that are later regimented by science or other academic disciplines. For example, *color* properties were known before they were classified into kinds and their determination dimensions discovered. We can find examples fitting this pattern in other domains, too. Various ethical properties—such as duties or justice—were probably known before any theory of duties or justice emerged. The theories that developed offered a classification of ethical kinds and their determination dimensions to fit the particular instances with which people were already acquainted. These are all cases in which methodological priority is given to properties.

Other times methodological priority lies with kinds. This is the case when a theorist hypothesizes some kind, along with specific or schematic determination dimensions, and then investigation pursues the properties belonging to the hypothesized kinds. When such properties are found, the hypothesis is confirmed. For example, Oscar Greenberg hypothesized the kind *color charge* for subatomic particles like quarks. This hypothesis was made without a prior acquaintance with such properties. Such a hypothesis was made, and then it was tested to determine whether there are instances of that kind.

3.3 How Many Kinds?

How many property spaces are there, and what are their determination dimensions? Recall that property and mere predicate spaces differ only in whether they succeed at picking out actual or physically possible properties. Predicates are easy to formulate, but the question here is how many of these predicates succeed in generating property spaces rather than mere predicate spaces. For example, are there property spaces corresponding to every physically possible predicate (i.e., predicate that can be instantiated without violating the laws of nature)? If so, then they are *abundant*. Or are there a rather limited number of property spaces—say, those few that carve nature at its joints? If so, then they are *sparse*. Such terms, of course, mark off only two possibilities along a continuum. This question—"How many?"—is one that any comprehensive property theory should address.

A fully satisfactory answer to this question would require staking out positions on the topics of meta-ontology as well as scientific or ontological reduction. But, without shouldering this task, this is a good place to note how, at least on the present framework, the "How many?" question is ambiguous. We should note what is *not* being asked by this question. The interesting question is not how many *properties* (abstract *particulars*) there are in our world, but how many *kinds* of properties there are in our world (and others). But an interesting ambiguity still remains: Does the "How many?" question concern super-determinate kinds only, or does it also include determinable kinds? Let us clearly separate these two different issues.

First, we might be asking how many property space *points* there are. Each such point represents a super-determinate kind. And to admit new super-determinate kinds is to admit new, physically possible particulars into one's ontology. This is because properties corresponding to this property space point are qualitatively (or causally, etc.) different from properties corresponding to any other. Typically, the "How many?" question is intended in this sense. For example, those who wonder if there is a property corresponding to every physically possible predicate are, in our terminology, wondering if there is a property space point (or region of points, if the predicate is determinable) corresponding to every physically possible predicate.

But the "How many?" question can also be understood in a second sense. One may wonder if, given all the property spaces, there are distinct kinds corresponding to each partitioning of these property spaces. Take, for example, the property space for *color*. Are there kinds corresponding to each way we can divide the colors—e.g., a kind corresponding to the conjunction of this shade of lavender, that range of oranges, and crimson? Unlike with the first interpretation, we already know that there are property space points corresponding to all these colors. Now we are asking if there are kinds corresponding to each such arbitrary division of property space points. This is another sense in which properties could be abundant.

Recall the assumption that kinds are theoretical constructs that represent possible particulars. Assuming that kinds are theoretical constructs, I am willing to be extremely permissive and allow for kinds corresponding to every partitioning of a property space. Of course, some partitions will be more natural, pragmatically beneficial, or theoretically beneficial than others. These partitions pick out the *more natural* kinds. As I see it, the second interpretation of the question does not raise an *ontological* issue, because it is already assumed that there are, or at least (physically) could be, properties corresponding to these property spaces. The ontological question arises with the first interpretation. And here we ought not to be so quick, as such carefree permissiveness in ontology is a sin against Ockham.

Without venturing into accepting an abundant conception of kinds, we also might wonder whether the individual determination dimensions that partially compose complex property spaces must themselves generate simple property spaces. That is, is there a simple property space corresponding to each determination dimension of a complex property space? For example, we have supposed that *color* is a three-dimensional determinable with hue, brightness, and saturation as determination dimensions. At issue here is whether *hue, brightness*, and *saturation* must themselves be simple determinables in their own right, distinct from their contribution to the property space of *color*.

I do not see a good reason for believing that this must hold as a general rule. Perhaps there must be such simple property spaces, simply due to the fact that there is a property space corresponding to every formulable predicate that is physically satisfiable. But I tend to think that the case for such simple property spaces should be made on an individual basis, requiring the same kind of support as in other cases. For example, there might be good scientific grounds for admitting an independent property space corresponding to the pitch determination dimension of *sound*. If kinds and their instances can earn their keep in virtue of making essential contributions to causal laws or generalizations, then *pitch* might do so. To borrow an example from Fred Dretske, the pitch of a sound may be that in virtue of which a glass shatters.[6] Pitch is just one of the determination dimensions of *sound*. But in this case it seems plausible to take the pitch of this sound as a property in its own right. This position is encouraged by the presence of a law or counterfactual dependency relating the pitch, specifically and "as such," to the shattering. The legitimacy of complex property spaces like *sound* could, similarly, be vindicated by laws relating the whole pitch-timbre-loudness bundle to some effect.

The status of complex property spaces is a difficult and deep philosophical problem for which I have no answer. For if there are complex determinables and kinds are not abundant, then why is it that some collections of determination dimensions combine to form kinds but others do not? And, why is it that some determination dimensions inevitably co-occur—e.g., hue, brightness, and saturation—to form complex kinds? Here we have a real unity. Perhaps the scientist can explain why the determination dimensions of certain complex determinables must co-occur, but this seems to be a problem ill-suited for the methods of philosophy.

3.4 Eliminativism about Determinables

The account of kinds I have offered assumes the legitimacy of determinable kinds and their instances. *Red* can be just as legitimate a kind as *scarlet*, and an object can instantiate each kind. I do not see any ontological

[6] Dretske (1988: 80).

competition arising between determinables and determinates, because the very same property is typically an instance of many determinates and determinables. Unlike other non-eliminativists about determinables such as Armstrong, Shoemaker, or Yablo, I then have no need to make a case for privileging determinable properties, in at least some contexts (e.g., due to proportionality considerations with respect to causation), over determinate properties.

Carl Gillett and Bradley Rives (2005) argue against determinable properties, making a case for a world of super-determinates only. In brief, their argument is that determinable properties do not contribute novel causal powers not contributed by their determinates. Further, they claim that ontological parsimony favors positing only one of these properties. Their conclusion is that we should not posit determinable properties *in addition to* determinate properties.

> Thus, the pessimist [eliminativist] concludes, we should not posit the existence of the determinable property H in addition to the determinate property G. And obviously similar conclusions hold for all the other determinates that are taken to fall under the determinable predicate 'H'. For in any situation where such a determinate is instantiated we can account for all of the relevant powers using instances of some determinate property...
>
> We may therefore summarize the idea behind such views as the claim that determinate properties are the only truthmakers for the statements involving both determinate and determinable properties.[7]

As a metaphysical point, they conclude that there are no determinable properties; instead there are only determinable predicates, which can be useful for explanatory purposes.

I accept much of the Gillett and Rives argument, such as the claim that we should not posit determinable properties in addition to determinate properties. I also accept that every property is super-determinate. But this does *not* mean that we should not posit determinable properties. We can admit determinable properties that are identical to determinate properties. This is not just a speculative assertion on my part. Rather, Chapter 2 presented a detailed account of determination, in terms of determination dimensions, which both distinguishes determination from realization and generates individuation conditions for properties. From these property space models and individuation conditions it follows that determinable

[7] Gillett and Rives (2005: 491, 492).

properties share instances with their determinates. A systematic, motivated, and plausible theory of kinds, properties, determination, and realization that has this consequence should be preferred to the distinctness of determinate properties and determinable properties (if such there were) that Gillett and Rives assume.

On the property identity account I favor for determinables and their determinates, there is something off about speaking, with philosophical seriousness, of *properties* as determinate or determinable. Because I do not think that the world is vague, I hold that every property is an instance of a super-determinate kind. But every property is simultaneously an instance of every determinable kind that super-determinate falls under. As was emphasized in Chapter 2 as the first feature of determination, it is a relation that holds between *kinds*. No property is *merely* a determinable any more than any concrete individual is vague. Properties are abstract, but they are specific. The determination dimensions of a kind provide us with the level of abstraction, but each property instantiates specific values along those dimensions. Determinables and their determinates share the same determination dimensions. Because even instances of determinables must instantiate specific determination dimension values, those "determinable properties" will simultaneously be "(super-)determinate properties." This is because those specific determination dimension values belong to the property spaces of those determinates as well.

To be clear, on my view instances of determinate kinds are identical to instances of their more determinable kinds. Instances of *scarlet* are identical to instances of *red* and *color*. I have simply *assumed* that every property is an instance of a super-determinate kind. As an armchair philosopher, I recognize that I am not in a particularly good position to defend this assumption that the world is some maximally specific way. The property theory I have defended, however, is not threatened by the possibility of *genuinely* determinable properties—that is, properties that are instances of determinable kinds without being instances of any more specific determinate of those determinable kinds (e.g., a red that is neither scarlet nor crimson and so on).

Jessica Wilson argues that both experience and science provide us with reasons to believe in such determinable properties.[8] She claims that the fact that we perceive objects as having determinable textures and shapes

[8] Wilson (2012: 5).

is one reason to posit determinable properties. Yet I fail to see how vague perception, any more than vague thought, justifies positing a world of determinable properties (so understood). Most of the other reasons she provides cite the fact that many scientific generalizations make reference to only determinable features—averages or ranges of values, say—and do not depend on the more specific values for mass or energy. But this is hardly a reason to think that there are determinable properties in the world, properties that lack more specific determination dimension values. Rather, it is just that those values are irrelevant for certain explanatory or predictive purposes. The error here is the same as if we were to reify the "average household," with its .94 children and a father with indeterminate hair color, because of the true generalizations we could make about it.

I have assumed that every property corresponds to a point in a property space (i.e., is an instance of a super-determinate kind), but if the world is mysteriously non-specific, then properties would correspond to larger regions of property spaces (e.g., the region for the most determinate determinable it satisfies). Such determinable properties would not be proven on armchair philosophical grounds—such as positing merely determinable *red* properties due to considerations about the metaphysics of causation. Rather, they would be more mysterious properties that genuinely are not accompanied by any corresponding determinate properties. In contrast, *red* properties are always accompanied by some instance of a more determinate shade. Genuine, merely determinable properties would have to be discovered by the empirical sciences or by the otherwise systematic study of the relevant conceptual scheme (mathematics, ethics, or whatever). But even for such properties the point remains that they have determination dimensions and they are simultaneously instances of every determinable they fall under—just as the supposed determinable *red* property would still be an instance of *color* as well. The subset model remains.

4

Multiple Realizability I: Its Role and Importance

Chapters 2 and 3 primarily dealt with determination, though there was some discussion of realization for contrastive purposes. In the next two chapters the focus turns to realization. In this chapter I offer a limited account of realization that is largely negative. Namely, it is a synchronic kind necessitation relation that is *not* determination. This account has practical payoffs when it comes to the issue of MR and its relation to reduction and autonomy in the special sciences. The main goal of these next two chapters is to argue for a constraint on accounts of MR—that it not be confused with multiple determination—that many have failed to meet. In this chapter I explain what I take to be some of the major problems for these unsuccessful accounts. In Chapter 5, I then offer my positive account of MR.

4.1 Sameness through Difference

The basic idea behind MR is sameness of kind through differences in the, typically lower-level, conditions that give rise to instances of that kind. The different conditions that give rise to instances of the very same kind are the multiple realizations. The classic example of a multiply realized kind comes from Putnam (1975: ch. 21). *Pain*, it was suggested, is multiply realized because human beings, alligators, and octopuses can have sameness of *pain* through differences in the neuroscientific kinds that give rise to these pains. These different neuroscientific kinds are the multiple realizations of *pain*.

Prior to appreciating the distinction between determination and realization one might have thought that, at a first pass at least, a multiply realizable kind is simply one that asymmetrically depends on other

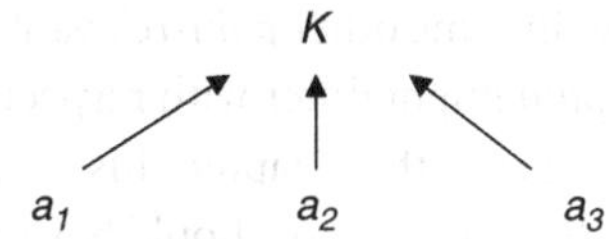

Figure 4.1 Asymmetric dependence

kinds. Figure 4.1 represents this position. Here *K* is supposed to be the multiply realizable kind, the '*a*'s the multiple realizations, and the arrows represent necessitation relations. On this model, *K* asymmetrically depends on the '*a*'s—i.e., each a_i kind-necessitates *K*, but *K* does not kind-necessitate any particular a_i. If MR were simple asymmetric dependence, then all kinds that can be modeled as such would provide us with examples of MR.

But recall that *red* and its various determinate shades also fit this model. If MR were simply asymmetric dependence, then *red* would have multiple realizations simply in virtue of coming in determinate shades. But this obviously is not the case. We do know that *red* cannot be reduced to *scarlet, crimson*, or *maroon*. And we do say that these are different ways (*scarlet, crimson*, and *maroon*) of being of the same kind (*red*). In some sense, this is sameness through difference. Such irreducibility claims and the "different ways" locution are familiar when speaking of MR. But we have also discovered that these are determination relations, *not* realization relations.

In contrast with the example of *red*, recall Figure 2.5 from Chapter 2 which depicts the relations between a super-determinate *pain* kind and its various neuroscientific realizations. Functionalists and others have argued, at least since Putnam, that members of different biological species—and even perhaps non-biological "species"—can instantiate the same *pain* kind while differing in the structural or compositional kinds that realize those pains. As with *red*, it could be said that each is a different way of being in pain and *pain* does not reduce to any one of these realization kinds. It might also be thought that just as *scarlet* and *crimson* differ along some of the dimensions by which colors are individuated, *alligator pain* and *human pain* differ along some of the dimensions by which pains are individuated. Let us assume, as before, that *pain* is a purely phenomenological kind. Again, nothing hinges on this assumption, but it makes exposition easier. Then just as *scarlet* and *crimson* are different shades of *red*, *alligator pain* and *human pain* might be different intensities of *pain* (or else they might differ in some other *pain*-relevant feature). That is, *alligator pain* and *human pain* might differ with respect to their "pain-ness." But, I will continue to argue in this chapter, this is not at all what proponents of the MR of *pain* have in mind or should have in mind. Rather, they intend to claim (or *should* intend to claim) that alligators and humans, say, can have exactly resembling pains, though differing in the conditions that give rise to their pains. Here I am not claiming that *pain* actually is multiply realized. Rather, I am simply clarifying what is required for *pain* to be multiply realized.

This is a much stronger sense of "sameness through difference" than what holds for the different shades of colors. Our distinction between determination and realization is intended to capture this difference. The multiple determinates of some kind *X* are the specific ways of being *X* that differ with respect to their *X*-ness. And a difference in *X*-ness is to be understood as a difference in the range of values along *X*'s determination dimensions. For example, *scarlet, crimson*, and *maroon* are three of the multiple determinates of *red* because they are specific ways of being red that differ with respect to their redness. And redness is understood in terms of a range of hue, brightness, and saturation values. Multiple determination is a type of asymmetric necessitation, but MR is an importantly different type of asymmetric necessitation. The multiple realizations of some kind *K* give rise to instances of *K* that do not *at all* differ in their *K*-ness. MR requires that there are at least some super-determinate kinds with distinct realization kinds.[1] Hence, multiple determination cannot be a species of MR, since the multiple determinates of some kind *K* differ in their *K*-ness and realizations of *K* are blind to *K*-ness. Nor can MR be a species of multiple determination, since the multiple realizations of a super-determinate kind *K* do not differ in their *K*-ness. They are not *K*-kinds at all, possessing distinct determination dimensions and non-determinable necessities that place them at another level of abstraction.

4.2 Realization

Chapter 2 advanced a theory of determination in terms of subsets of determination dimension values. But there are other interesting kind

[1] Haug (2010) challenges this claim, however. His objection is addressed in §5.2.

necessitation relations as well, and realization is prominent among them. Realization first gained prominence as the key concept in understanding the relationship between abstract computational or functional states and the hardware that implements them. The concept was then generalized to pick out the relationship that holds among various higher-level kinds and the lower-level kinds in virtue of which they obtain. This generalized concept of realization eventually became the critical concept in formulating the thesis of physicalism.[2] The concept of realization was long overshadowed, however, by one of its offspring concepts—MR. It is unfortunate that MR was long discussed in the philosophy of mind without many of the disputants having, or at least advancing, any clear understanding of the realization relation itself. But, eventually, two general approaches to the metaphysics of realization rose to prominence.

On the first approach, functional kinds, the prototypes of multiply realizable kinds, are taken to be *second-order kinds*. According to this approach, an object instantiates a functional kind simply by instantiating some first-order kind or other that meets a given functional specification. The essence of the functional kind is supposed to be exhausted by this quantification over first-order kinds that satisfy a functional specification. And the first-order kinds that meet the functional specification are supposed to be the realizations. More generally, the realization relation is that relation which holds between such first-order and second-order kinds. This account is endorsed by Kim (1993b, 1998, 2005), but traces back to David Lewis (1983: ch. 6) and others.

The second approach, while compatible with the first, focuses on the causal powers of the realized and realizing kinds. On this view realization is understood as *inheritance of causal powers*. Namely, realized kinds are those kinds that inherit a proper subset of the causal powers of other kinds—their realizations. Advocates of this approach also typically hold that the essence of a kind lies with its causal powers. This approach is most prominent in Sydney Shoemaker (2001), and is also found in Kim (1993b) and Carl Gillett (2002, 2003).

There are serious problems with both approaches, however. First, consider the approach that takes realization to be the relation that holds between first-order and second-order kinds. On this account, the instantiation of a multiply realizable kind, say *pain*, on a particular occasion is

[2] See, for example, Kim (1998) and Shoemaker (2001).

simply identical to the instantiation of some specific realization kind, say NS_1. But we have already argued that multiply realized kinds have different determination dimensions than do their realization kinds. If this is correct, then an instance of *pain*, on the assumption that it is multiply realizable, will always be distinct from an instance of the realization kind, say NS_1, in virtue of which it obtains. This is because they have distinct determination dimension values (and perhaps non-determinable necessities as well), so their instances will not exactly resemble one another (let alone be identical). Speaking more generally, multiply realizable kinds cannot be identified with second-order kinds ranging over their first-order realizations, because the multiply realizable kinds and their realizations have distinct determination dimensions.

This approach also has the more basic flaw that being first-order or second-order is not part of the nature of any kind, but only our characterization of it. It does seem that some kinds, such as functional kinds, are more naturally characterized as second-order than are others. However, any kind can also be given a second-order formulation. For example, we can characterize *red* as second-order relative to its various (first-order) determinate shades. In such a case we characterize *red* as the kind that is instantiated whenever, and only whenever, some specified first-order kind or other (specified functionally, if this is possible for colors, otherwise by simple enumeration or some other method) is instantiated. But, again, this is not a case of realization—e.g., *scarlet* and *crimson* are not two realizations of *red*. So, we can appropriately characterize kinds as standing in such second-order to first-order relations even when no realization relation holds among them. The realization relation must be something else.

Another potential problem with the second-order approach is that it risks making multiply realizable kinds causally irrelevant. An instance of a second-order kind is always identical to an instance of the first-order kind that realizes it on that occasion. This identity means that second-order properties (i.e., instances of second-order kinds) do not contribute any new causal powers that are not already found with the first-order properties (i.e., instances of first-order kinds), as they simply *are* first-order properties.[3] This is a problem for those who think that instances of multiply realized kinds *do* contribute causal powers that are distinct from those

[3] See Kim (1993a: part II), Kim (1998), and Kim (2005).

found with their realizations. (Of course, others will see this as a handy way of saving mental causation, by reducing it to the causal efficacy of the lower-order kinds.)

Next consider the view that realization is to be understood in terms of inheritance of causal powers. This view is problematic because such inheritance is neither necessary nor sufficient for realization. First, let's consider sufficiency. If the causal powers of realized kinds were inherited from their realizations, it would seem at least as plausible that the causal powers of multiply determined kinds would be inherited from their determinates. The causal powers of *red*, let us suppose, are a proper subset of the causal powers of its various determinate kinds *scarlet, crimson*, etc. However, *scarlet* and *crimson* are not distinct realizations of *red*. Rather, they are distinct determinates of *red*. So, inheritance of causal powers is not sufficient for realization.

Nor is such inheritance of causal powers necessary for realization. Realized kinds often possess causal powers that are not inherited from (i.e., a proper subset of the causal powers of) their realization kinds. For example, *transparent* is a kind that confers causal powers to the objects that instantiate it, such as the power to let light pass through them, which need not be found with its realization kinds *even taken collectively*. This is especially obvious if, as I think we should, we tie a kind's causal powers to its non-determinable necessities and determination dimensions, for these dimensions are distinct whenever the realization relation obtains. Causation depends on certain relations—e.g., law-like connections, counterfactual dependencies, or some simple primitive causal relation—holding among abstractions. The present theory of abstractions explains how realized and realization kinds provide us with different levels of abstraction (i.e., determination dimensions and non-determinable necessities) and, therefore, different causal powers. None of this is to deny that the causal powers of realized kinds depend on the causal powers of their realization kinds. A realization base of a kind might synchronically necessitate the higher-level kind's causal powers, as I assume is typically the case, without those causal powers also belonging to the realization base itself.

These approaches fail in that they construe realization both too broadly and too narrowly. They are too broad in that they mistakenly include determinates as realizations. And they are too narrow as well. Some realized kinds (such as functional and dispositional kinds) might be appropriately or at least naturally characterized as second order, and some realized kinds might be such that they inherit all their causal powers from their realizations, but this is not true of all realized kinds. Realized kinds can be non-functional, and they certainly can be distinguished along determination dimensions distinct from those of their realizations. So, they do not merely possess a subset of the essential features or causal powers of their realizations.

I have objected to two prominent accounts of realization and shown that the present account, which contrasts realization with determination, requires an alternative understanding of the relation. So if not in these two ways, how should realization be understood? I am not sure that any one answer will suffice, as distinct realization concepts are sometimes lumped together.

One major division in the use of 'realization,' as discussed within the philosophy of mind, concerns the nature of the realized states. This difference in realized states might mark a difference in realization relations well.

1. *Abstract or computational state realization$_1$ in the physical.* For example, a given tape and reader/writer realizes$_1$ a Turing machine. Or, my computer realizes$_1$ a particular word-processing program.
2. *Higher-level physical kind realization$_2$ in lower-level physical kinds.* For example, a certain microphysical structure realizes$_2$ the transparency of a window.

One might think that each of these is an example of *the* realization relation of interest to philosophers of mind. After all, different philosophers of mind have endorsed each model—Turing machine functionalism and higher-level state accounts—as apparently competing accounts of how the mental is realized in the physical. But closer inspection reveals that distinct relations are likely at work in these examples.

Realization$_1$ relates something non-physical, such as abstract or computational states, to something physical.[4] Further, the existence of such abstract or computational states in no way depends on their physical realizations. Turing machines are non-spatio-temporal, mathematical entities that are distinct from their realizations and are in no way ontologically dependent upon them. In this regard, realization$_1$ contrasts sharply with realization$_2$. The transparency of a window is a physical property that

[4] Here the word 'abstract' is used not in our original sense, but instead to refer to something non-spatio-temporal.

would not exist but for its lower-level realizations. And both transparency and its realizations are physical phenomena. These differences, I believe, are sufficient to establish that realization$_1$ and realization$_2$ are distinct relations. Whenever realization$_2$ obtains, the instances of the realized kinds are present with their realizations and ontologically dependent on them—e.g., the transparency of the window would not exist were it not for its realizations. Realization$_2$ is a relation of ontological dependence. But this is not the case with realization$_1$. Turing machines and other abstract or computational entities are not ontologically dependent on the devices that are their realizations$_1$.

These are both legitimate relations worth investigating. However, realization$_2$ is of greater relevance to the problems of present interest, namely the relationships among the kinds posited by the special sciences and basic science. Instances of, say, psychological and biological kinds are present with and ontologically dependent on the chemical kinds that are their realizations. Realization$_2$, to use the language of §2.1, is an intimate kind necessitation relation. Surely this is a better way to think of mind–brain relations, rather than in terms of realization$_1$. After all, mental states obtain in virtue of brain states and are present with them. More generally, realization$_2$ is the relevant relation for characterizing the familiar "levels" or "layered" hierarchical picture of the sciences. For these reasons, I will be concerned with explicating and analyzing realization$_2$, and all future talk of realization should be understood as referring to this relation.

In addition to this major division, there are other dimensions along which concepts of realization differ. For example, Shoemaker (1981) distinguishes *core realizations* and *total realizations*. I accept the spirit of his distinction, as well as his terminology, but I will use the terms in slightly different senses. For Shoemaker, the distinction between core realizations and total realizations is made in terms of which functional properties Ramsey–Lewis sentences are utilized to express.[5] But the spirit of the distinction can be generalized so that it applies to non-functional properties as well.

I think that it will prove helpful and productive to think of core realizations as those that make the most significant, or at least salient, contributions to bringing about their realized kinds. As the significance and

[5] Shoemaker (2003: 265).

salience of such contributions admit of degree, core realization itself admits of degree (i.e., something can be a more or less core realization). The extent to which a property contributes to realizing a kind is, in at least some cases, a fairly objective matter. For example, given the central role that the heart plays in circulating blood, it is an objective fact that the heart is a core realization of the circulatory system. This is perfectly consistent with there being other core realizations of the circulatory system—e.g., veins and arteries—as well as background conditions that need to be met in order for the system to function properly or at all. We can also understand something to be a core realization due to the salience of its contribution to its realized kind. Salience is relative to both observers and their interests. Some property of a system might not make a major contribution to realizing some kind in an objective sense, but nevertheless that property makes an essential contribution. If this property is rare, easily manipulated, or otherwise garners special interest, then it could be salient and thereby be a core realization. So, something can be a core realization for either objective, metaphysical reasons or for epistemic, explanatory reasons. This is analogous to a phenomenon we discover when analyzing the concept of causation. We can judge some condition as *the* cause of an effect either because that condition made the greatest contribution to bringing about the effect or because that condition simply was salient.

Core realizations are not strictly sufficient for their realized kinds. Instead, core realizations must be supplemented with other properties of the relevant system or placed in the appropriate context in order to successfully bring about those realized kinds. Total realizations, which are strictly sufficient for their realized kinds, are combinations of core realizations and these supplemental conditions. For example, the total realization of the human circulatory system includes all the components, besides just the heart, that are individually necessary and jointly sufficient for circulating blood. Presumably, these supplemental conditions are internal to the human body. But other times the supplemental conditions extend beyond the system that instantiates the realized kinds.[6] For example, a certain piece of paper is the core realization of a US dollar. But

[6] My usage differs from Shoemaker's in this aspect. Shoemaker's total realizations are all properties of the same system, whereas on my usage total realizations include background or contextual conditions that extend beyond that system. This usage also contrasts with that of Wilson (2001).

that piece of paper must be properly situated within a larger institutional setting—involving the US Treasury, etc.—in order for that piece of paper to be worth one dollar. These supplemental conditions, extending into the past as well as into US Treasury buildings, are part of the total realization of the dollar bill. It is widely thought that content-bearing mental states, such as beliefs and desires, are like dollar bills in this regard. The core realizations of mental states are states of the central nervous system, but it is widely accepted that the content of many such states is partially determined by the external world.

The distinction between core and total realizations is important to keep in mind. And we should recognize that there are realization relations of each type. I will offer formulations of each realization relation. We can leave it to the various scientific and philosophical disciplines to determine whether the more interesting MR claims cover core or total realizations.

Undoubtedly there are other ways in which the concept of realization could be sensibly fragmented. Fortunately, I will not be concerned with giving a precise account of *the* realization relation. However it is to be understood, realization should be contrasted with determination for the reasons already given, as well as those developed later in this chapter. I aim to provide the most general characterization of realization that covers the primary examples of scientific and philosophical interest. I think that most will accept the following three data points as offering plausible restrictions on realization:

1. Realizing properties are *sufficient*, in some sense or other, for the properties they realize. The realization properties might strictly necessitate the realized property (in which case they form the total realization), or they might necessitate the realized kind only in certain contexts or given certain background assumptions (in which case they form a core realization).
2. Realized properties are *present with* (co-located with, constituted by, etc.) their realizing properties.
3. Realized properties *depend* on their realizing properties, in either an ontological or explanatory sense.

These features are widely accepted, though there have been a few dissenters.[7] And these features are accepted for good reason. Realization has been

[7] Wilson (2001), for example, argues that 1 and 2 are not jointly satisfied by many realizations. His point, I think, is largely correct. Realized properties are sometimes only "present

invoked to explain mental–physical relations and higher-level/lower-level relations more generally. The realization relation is invoked by physicalists to express the idea that particular physical properties are sufficient for mental properties, mental properties are present with such physical properties, and the mental depends on the physical. A satisfactory account of realization, then, will capture these three features of the relation. But do note that these three features are also true of the determination relation as well. For example, *scarlet* properties are sufficient for *red* properties, are present with them, and explain their occurrence (though this last claim is more controversial). So we must add something to distinguish realization from determination.

4. The kinds to which realization properties belong are not determinates of any of the kinds to which their realized kinds belong.

I have proposed that the key difference between realization and determination is that realized and realization properties have distinct determination dimensions, whereas determinables and their determinates do not have distinct determination dimensions. This was one of the main lessons learned from Chapter 2.

I then offer the following conditions to meet our four data points:

Necessary and Sufficient Conditions for Total Realization

Kinds B_1–B_n collectively totally realize kind *A* if and only if:

1. B_1–B_n collectively form a minimal set[8] of *B*-kinds that synchronically kind-necessitate *A*;
2. *A*-properties obtain in virtue of B_1–B_n properties (collectively), whenever the latter obtain;
3. B_1–B_n and *A* possess different determination dimensions; and

with" their core realizations, though these core realizations are not sufficient for their realizations. Such is the case with the dollar bill. I take this point into account in formulating distinct accounts of total and core realizations.

[8] The "minimal set" qualification is intended to rule out any proper subset of B_1–B_n kind-necessitating *A*. Other, and perhaps smaller, sets of *B*-kinds might also kind-necessitate *A*, however.

4. *A*-properties are co-located, by necessity, with some core (if not the entire collective) of B_1–B_n properties whenever kinds B_1–B_n are collectively instantiated.[9]

For example, suppose that it is a metaphysical truth that various neuroscientific kinds NS_1–NS_{12} synchronically kind-necessitate *pain*. Whenever neuroscientific kinds NS_1–NS_{12} obtain, *pain* obtains in virtue of the instantiation of the neuroscientific properties. Further suppose that the determination dimensions for *pain* differ from the determination dimensions for these neuroscientific kinds. Finally, whenever kinds NS_1–NS_{12} are collectively instantiated, a *pain* property is co-located with some core (if not all) of those *NS*-properties. In such a case, neuroscientific kinds NS_1–NS_{12} totally realize *pain*. In fact, they totally realize some super-determinate *pain* kind as well as every determinable that *pain* kind falls under.

Some further comments on each of these conditions are in order to clarify this analysis. First, I assume that it will often, if not typically, be the case that many *B*-kinds collectively realize a single kind, *A*. In our last example, twelve neuroscientific kinds collectively realize a single *pain* kind. However, it is also possible that $n=1$. That is, it is possible that a single *B*-kind kind-necessitates *A* all by itself. Note that such situations need not be situations in which kind *A* is singularly, as opposed to multiply, realized. Perhaps other *B*-kinds also individually kind-necessitate *A*, or collections of other *B*-kinds collectively kind-necessitate *A*, or kinds from distinct conceptual schemes (*C*-kinds, *D*-kinds, etc.) also kind-necessitate *A*.

Second, note that conditions 1 and 4 appeal to some sense of necessitation. But is this metaphysical necessitation, nomological necessitation, or some other variety? I understand realization to be a relation of metaphysical necessity. For one, I think that this is true of the canonical examples of realization—e.g., H_2O and XYZ metaphysically necessitate the water of their respective worlds. It also better accords with our understanding of physicalism in terms of both realization and metaphysical necessity. For example, if the brain merely nomologically necessitated phenomenal consciousness in the manner David Chalmers has argued,[10] then the brain

[9] Realization differs from determination in that it is limited to spatio-temporal kinds. All realized and realization properties are spatio-temporal, given the present focus on realization$_2$, though not all determinable or determinate properties are spatio-temporal.

[10] Chalmers (1996, 2010).

does not realize phenomenal consciousness and physicalism is not true. Nevertheless, others might prefer that realization be understood in terms of the less restrictive nomological necessity. We could speak of either variety of the realization relation, of course, and the account can be easily modified.

Third, it would be helpful to have some further understanding of the "in virtue of" qualification found in condition 2. This "in virtue of" locution could be taken as a primitive and metaphysically fundamental, but various ways of cashing out this expression are available. In our example, this qualification could be met by the fact that the neuroscientific properties *explain* the presence of pain, or that the necessitation is *asymmetric*, or that neuroscientific properties are *lower-level* than their co-located pains. I prefer to understand this locution in terms of Robert Cummins' notion of system or property analysis.[11] The idea is that some system's or object's possession of a property can be analyzed in terms of that system's or object's component parts and their organization. It is natural to describe these component parts as mechanisms, and I am largely sympathetic to the development of Cummins's theory by Carl Craver (2001) in mechanistic terms. Part of this analysis involves the discovery of kind necessitation relations, like those identified in condition 1. But realization requires a further metaphysical or explanatory grounding for these relations. And this is what the "in virtue of" clause is supposed to provide. A system or object's possession of an *A*-property obtains because that system or object possesses *B*-properties arranged in a certain way. This "because" can be taken as primitive or it could be understood as shorthand for the explanatory standards accepted by successful scientific practice. Typically, if not always, the "in virtue of" direction will be upward. That is, higher-level properties obtain in virtue of the arrangement of lower-level properties. But I do not want my account of realization to demand this. It is at least epistemically possible that the converse could hold: lower-level properties obtain in virtue of the arrangement of higher-level properties. This would be appropriately described as a case of downward realization.

Fourth, it might be a bit awkward to say that *A* possesses different determination dimensions than does B_1–B_n, give n that *A* has one set of determination dimensions and there are different determination dimensions

[11] Cummins (1983: 15).

for each of the *B*-kinds B_1–B_n. But the idea is that the *B*-kinds B_1–B_n and *A* are at different "levels of abstraction." And this is to be understood as follows: *A*'s determination dimensions are not identical to any individual *B*-kind's determination dimensions nor are *A*'s determination dimensions the conjunction of B_1–B_n's determination dimensions or otherwise derivative from them.

Fifth, I have promised to offer nominalistic versions of my central claims and analyses. But, condition 4 is phrased in terms of Property Realism—properties can be co-located with one another only if properties actually exist. So, it would be beneficial to have a nominalist-friendly substitution for condition 4 that captures the spirit of this analysis. The general idea behind condition 4 is that, to stick with our example, the pain is simply there with the neuroscientific substratum that realizes it. Whether or not there are properties, we can certainly make sense of spatio-temporally locating these pains and neuroscientific substrata, however difficult and vague that might be in practice. If anything, *pain* is one of the more difficult cases in this regard—most other scientifically investigated states are easier to place. The nominalist should simply understand condition 4 as stating that whatever spatio-temporal location can be assigned to these neuroscientific substrates can also be assigned to these pains, and so too for any other realized kind. Even though the nominalist denies the existence of properties, she often understands where they would be located were they to exist.

Finally, condition 4 contains the notion of a core group of realizing properties. This should be understood in the previously articulated sense of core realizations, in contrast with total realizations. I recognize that others will prefer to think of realization as the relation that holds simply between these core realizations and the realized kind. And I offer the following conditions to capture their preferred variety of realization:

Necessary and Sufficient Conditions for Core Realization

Kinds B_1–B_n collectively core realize kind *A* if and only if:

1. B_1–B_n collectively form a minimal set of *B*-kinds that, in contexts C_1–C_m, synchronically kind-necessitate *A*;
2. *A*-properties partially obtain in virtue of B_1–B_n properties (collectively), whenever the latter obtain,
3. B_1–B_n and *A* possess different determination dimensions; and

4. A-properties are at least partially co-located, by necessity, with B_1–B_n properties whenever kinds B_1–B_n are collectively instantiated in contexts C_1–C_m.

The core realization analysis differs from the total realization analysis with respect to conditions 1, 2, and 4. Let's begin with condition 1. Suppose that neuroscientific kinds NS_{42}–NS_{73}, in some possible Earthly context, form a minimal set that collectively kind-necessitates the mental kind *believes that Aristotle was Plato's most famous student*. Then, kinds NS_{42}–NS_{73} meet the requirement given in condition 1, relative to this supposed Earthly context. This context likely includes facts about the initial baptisms of the names 'Aristotle' and 'Plato,' the causal history of how these names were acquired by our believer, various other facts about the community of speakers, and perhaps even factors like the rest of the person's neurophysiology and behavioral history. (All of this detail should also make the understanding of realization in terms of metaphysical necessity more palatable.) These conditions are part of the total realization of the belief kind, because the core realizations would generate beliefs with different contents in different contexts (or could fail to generate a belief at all).

The second difference concerns condition 2, which now includes the qualification 'partially.' The justification for this qualification should be obvious. The realized kind A obtains in virtue of both B_1–B_n and C_1–C_m. Without the latter properties or contexts, there would be no A-property. So it is only partially in virtue of the B-properties that the A-property obtains.

The third difference is with condition 4. The (core) realized kind need not be wholly co-located with B_1–B_n, but it must be at least partially co-located with B_1–B_n. With core realization, the realized kind can also be present beyond the core realizations. This is especially obvious when we recognize that some core realizations are such for purely epistemic or explanatory reasons. The situation here is the opposite as for total realizations. A realized kind might be present with only a subset of its total realization. For example, I presume that the belief that Aristotle was Plato's most famous student is not present with the linguistic and non-linguistic conditions that make up the relevant context in our example. Instead, the belief is located just with the agent. With core realizations the situation is reversed. A realized kind might be present beyond some of its core realizations—e.g., recall the example of the circulatory system and its various core realizations.

These are two ways of understanding what it is for one *kind* to realize another kind. This parallels our understanding of the determination relation, which we also took as holding among kinds. But we can also speak of properties, the particular things out there in the spatio-temporal world, as realizing other properties. And properties, just like their kinds, can realize other properties in either the total or core senses. Property-level realization occurs whenever there are instantiations of the B_1–B_n kinds, as well as an instantiation of the *A*-kind, and the four conditions for either total or core realization are met. In particular, the *B*-properties and *A*-property must be co-located in the manner specified by condition 4.

Note that there is no need for a property version of determination. Whereas one property can realize another property, it is not possible for one property to determine *another* property. As was argued in Chapter 2, every property is an instance of some super-determinate kind as well as being an instance of every determinable that super-determinate kind falls under. So, an identity always holds between the instance of a super-determinate kind and the co-located instances of all the determinables that super-determinate falls under.

I have now offered detailed conditions for two varieties of realization. Each variety of realization—total and core—has its own utility in theorizing. But I imagine that the core realization relation will be more appropriate for many theoretical purposes. This is because total realizations, in order to meet their strict standard of sufficiency, must incorporate many factors that are simply assumed as background conditions in real-world situations. Scientists have limited interests, energy, and time. Consequently, the practitioners of any scientific discipline will assume that some parameters have already been fixed. In other words, these distinct scientific disciplines tacitly assume their own conditions C_1–C_m. These practitioners often focus on discovering the mechanisms and principles of direct or proximal relevance to the phenomena within their domain. Decompositional functionalism is a very clear example of a scientific approach that is primarily focused on finding such core realizations.

Let us suppose that some neuroscientific kind NS_1 is (synchronically) sufficient, in a given context, for *pain*. Further suppose that NS_1 has determination dimensions distinct from those of *pain*, determination dimensions that are relevant to neuroscientific classifications and theorizing. These determination dimensions might concern things like neural structures and signals. The determination dimensions of NS_1 are assumed to be irrelevant to individuating *pain*. Pains, let us suppose, are individuated on purely phenomenological grounds, regardless of their neuroscientific grounding. Nevertheless, NS_1 gives rise to *pain*. In short, NS_1 is a realization of *pain*. A realization of *pain* will (at least partially) give rise to the phenomenology distinctive of *pain*, though possessing determination dimensions that are irrelevant to individuating *pain*.

It might turn out that there are interesting correlations between neuroscientific kinds and *pain* kinds. Of course, this is inevitable if certain neuroscientific kinds realize, and hence necessitate, *pain* kinds. If this is true, then why won't the determination dimensions of *pain*'s realizations be relevant to individuating pains? The answer here is subtle but important. Though it is possible for the realizations' determination dimension values to direct us to a *pain* classification, it is not *because* pains have realizations with those dimension values that they are so classified. Rather, in the supposed case, the realizations' determination dimension values simply correlate with the determination dimension values in virtue of which the *pain* classification obtains (i.e., *pain*'s determination dimension values).

It is often the case that changing the determination dimension values of a realization kind results in a change in determination dimension values for the realized kind. For example, a change in the neuroscientific conditions that give rise to *pain* might result in a change in *pain* intensity. In fact, I assume this is commonplace. But, significantly, in such cases there need not be a one-to-one correlation between realization kinds and determinate kinds. In our example above, NS_1 not only gives rise to the determinable kind *pain*, it also gives rise to some super-determinate kind of *pain*. Significantly, it is possible that distinct realization kinds can give rise to the same super-determinate kind—as I argue in Chapter 5, this is the very possibility of MR. For example, it is at least epistemically possible that distinct neuroscientific kinds NS_1 and NS_2 give rise to the same super-determinate kind of *pain*.

It should now be clear that determination and realization are mutually exclusive, as the determination relation relates kinds that necessarily share the same individuation conditions (determination dimensions and non-determinable necessities), whereas realization relates kinds that necessarily do not share the same individuation conditions. Of course, we could use the term 'realization' in different senses. And we could concern ourselves with various other synchronic kind necessitation relations as well. But I will argue that the distinction between determination and realization, as made here, makes an essential contribution to understanding a concept of MR that is both very interesting and fruitful. And I also think that realization so understood is the concept that many philosophers of mind have been employing all along, though they have not explicitly characterized it in this manner.

4.3 Multiple Realizability and Autonomy

With these comments about realization out of the way, we can now return to the topic of MR. MR gained prominence, originally through the work of Hilary Putnam and Jerry Fodor in the 1960s, as the key concept (along with Donald Davidson's anomalism of the mental) in establishing non-reductionism and autonomy as the dominant position in the philosophy of mind. It soon was widely accepted that the special sciences more generally are replete with multiply realized kinds and that the MR of a kind was sufficient to establish the autonomy of that kind. I believe that the primary importance of MR still lies in this connection. So, in this section I will further characterize the role that MR is supposed to play in various arguments for the autonomy of the special sciences. I then claim that, whether or not these arguments are sound, it is a significant virtue of an account of MR that it shows how MR plays the role asked of it in the premises of the arguments provided below. In Chapter 5, I argue that the account of MR that I offer possesses this virtue. I acknowledge that others possess different concepts that they label 'multiple realizability' and might, initially at least, judge the present proposal to be off-target. But I am at least offering an account of a philosophically interesting concept of MR. Further, this is the concept that most heavily bears on issues of autonomy and reduction, the reasons why the term was introduced in the first place.

Despite Putnam and Fodor's best efforts (and those of many others), reductionism still has its defenders. And many have denied that MR entails non-reductionism. I often find these debates between professed reductionists and non-reductionists frustrating, as the term 'reduction' is frequently used in various senses by its participants. I then get the sense that those who adamantly disagree with one another on such matters often are merely engaged in a verbal dispute. That is, they are just using the word 'reduction' in various senses, and if they simply agreed on a fixed usage they would find that they do not have a substantive disagreement. I do acknowledge, though, that there can be quite significant disputes over which sense of 'reduction' is most theoretically and practically beneficial and, therefore, how best to define or understand reduction. As I see it, reduction has been used to capture a variety of ontological, explanatory, and methodological claims. I therefore recommend that we abandon the ambiguous word 'reduction' and instead replace it with direct reference to these more specific ontological, explanatory, and methodological claims. So instead of considering the consequences that MR has for reduction *simpliciter*, I will consider specific and plausible ontological, explanatory, and methodological consequences of MR. There are various senses of 'autonomy' standing in contrast to each of these various senses of 'reduction.'

The arguments I offer here are both simplistic and schematic. Undoubtedly, they could be further developed by others. I aim to sketch the theoretical work that MR is widely supposed to perform. Each of the three arguments that I offer contains two premises and a conclusion. The second premise of each argument is a claim that some kind is multiply realized. Such premises will simply be treated as assumptions. The first premise of each argument connects the MR claim to some conclusion concerning the autonomy of those kinds. I will argue that these first premises are plausible and that the present understanding of realization meets the needs to which the concept is applied in such premises. In §4.4 I point out where some alternative accounts of realization fail on this score. In the present section I offer only brief presentations of the arguments. I elaborate on and defend their respective first premises in §5.4.

4.3.1 *Ontological Autonomy*

MR was originally introduced as the key concept in a major objection to the type-identity theories of mind that rose to prominence in the 1950s. Most prominent among these were the theories of U. T. Place (1956), Herbert Feigl (1958), and J. J. C. Smart (1959). They offered ontological theories according to which mental kinds are identical to physical kinds. These kind identities were supposed to be like other scientific identities that were discovered *a posteriori* and, therefore, assumed to be contingent—e.g., that lightning is identical to motion of electric charges. Likewise, to use a common fictional example, the mental kind *pain* was supposed to be identical to some physical kind, such as *C-fiber firing*. This was perceived to be a scientifically respectable solution to the mind–body problem in that it offered a reduction analogous to those found in other scientific domains.

But this proposed solution, elegant as it was, proved to be flawed. Putting aside Kripkean objections to contingent theoretical identities, the main problem with this proposal is that the required mental–physical correlations simply fail to hold. It is a simple truth about identity that if $a = b$, then a is present only when b is present (and vice versa). So if *pain* is identical to *C-fiber firing* (or some other physical kind candidate), then pain is present only when C-fiber firing (or some other physical kind candidate) is present. But according to the MR objection to mental–physical kind identities, no such correlations hold between *pain* (or other mental kinds) and physical kinds. Putnam vividly presented this objection as follows:

> Consider what the brain-state theorist has to do to make good his claims. He has to specify a physical-chemical state such that any organism (not just a mammal) is in pain if and only if (a) it possesses a brain of a suitable physical-chemical structure; and (b) its brain is in that physical-chemical state. This means that the physical-chemical state in question must be a possible state of a mammalian brain, a reptilian brain, a mollusc's brain (octopuses are mollusca, and certainly feel pain), etc. At the same time, it must not be a possible brain of any physically possible creature that cannot feel pain.[12]

From this specific objection we can extract a more general argument schema for the ontological autonomy of kinds:

MR Argument for Ontological Autonomy

OP1. Whenever some kind K can be realized by various S-kinds or collections of S-kinds, S_1–S_n (where S is a science or conceptual scheme), kind K is not identical to any single S-kind or collection of S-kinds, S_1–S_n. Nor is K identical to the disjunction of S_1–S_n. In such cases K does not ontologically reduce to any S-kind or collection of S-kinds, S_1–S_n. Instead, kind K is ontologically autonomous with respect to S-kinds S_1–S_n.[13]

[12] Putnam (1975: 436).

[13] There is an important difference between the use of 'S_1–S_n' here and the use of 'B_1–B_n' in the analyses of realization. Here, every S is supposed to be a realization (either total or core) of some kind K all by itself. For this reason, individual 'S's can be collections of S-kinds. In contrast, the kinds B_1–B_n in the analyses of realization were supposed to collectively realize

OP2. Kind *A* can be realized in various *S*-kinds or collections of *S*-kinds, S_1–S_n.

OC. Therefore, kind *A* does not ontologically reduce to *S*-kinds S_1–S_n. Instead, kind *A* is ontologically autonomous with respect to *S*-kinds S_1–S_n.

OP2, I take it, is an MR claim. OP1 connects such MR to kind-level ontological non-reductionism and autonomy. In short, this argument points out that identity is a one-to-one relation, though MR is one-to-many. So, *A*'s MR with respect to some kinds is incompatible with *A*'s being identical to or ontologically reducing to those kinds. I allege that it is a virtue of an account of MR that it explains why MR is connected to ontological non-reductionism and autonomy in the way offered by OP1.

This argument schema purports to establish that MR entails kind-level ontological autonomy. Two points of clarification should be made. First, this argument does not directly establish that multiply realized *properties*, understood as particulars, are ontologically autonomous with respect to their various realization properties. There might be a different argument or reason for accepting such token-level ontological autonomy, though. Indeed, such an argument was presented in Chapter 2 to justify the property individuation conditions offered there. However, functionalism and other non-reductive accounts of the mental that countenance widespread MR have long been thought to be compatible with mental–physical token identities. The relevant tokens have typically been taken to be events, but property instances or states of affairs have also been candidates. According to these popular non-reductive theories there is a kind-level pluralism but a token-level monism. So we can distinguish between weaker (kind-level only) and stronger (kind-level plus token-level) versions of ontological autonomy. Here I am presenting an argument schema only for weak ontological autonomy.

Second, as I am understanding kind-level ontological autonomy, a kind *A* can be autonomous with respect to kind *B* even though *B* kind-necessitates *A*. Kind-level ontological autonomy simply requires kind distinctness and the inability to connect the kinds by simple logical operations like conjunction or disjunction. There is a straightforward

some kind *A*. So, to clarify, each *S* is either a single *S*-kind or a collection of *S*-kinds that, on its own, serves as a total or core realization of *K*.

ontological understanding of such distinctness for Kind Realists. But I have endorsed Kind Nominalism. So on my view kinds are not, strictly speaking, ever ontologically autonomous since no kinds exist. Nevertheless, we can take our bookkeeping seriously—as scientists are prone to do—and insist that there is a clear sense in which kinds can be said to be distinct even if, strictly speaking, no kinds exist. Compare how a metaphysician, speaking with the vulgar, can acknowledge a sense in which some ordinary objects are distinct even though, strictly speaking, she denies the ontological reality of ordinary objects. But further, I have argued that properties belonging to multiply realized kinds are ontologically autonomous with respect to properties belonging to their realization kinds. And this property-level claim is often judged to be a stronger ontological autonomy claim than is the claim for the kind level. Kind-level autonomy can then be derived from property-level autonomy. For these reasons, even though I do not endorse Kind Realism, there is a clear sense in which I have connected MR with kind-level ontological autonomy.

4.3.2 *Explanatory Autonomy*

If MR entails kind-level ontological autonomy, then it might be expected that the theoretical work (e.g., explanations, predictions, and generalizations) done by predicates invoking MR kinds is distinct from the theoretical work done by predicates invoking their realization kinds. For if kinds "crosscut" one another, as is the case with multiply realized kinds and their realizations, why think that the laws and patterns captured by the vocabulary for the one kind can be mirrored by using the vocabulary for the other kind? Many have argued that higher-level sciences, in general, discover patterns that cannot be captured by the vocabulary of their lower-level realization sciences.[14] This is because the lower-level detail is simply irrelevant to the recognition of the higher-level pattern, or it actually gets in the way of our understanding the phenomenon at hand. So, higher-level sciences are explanatorily autonomous. Once again we can turn to Putnam as a chief representative of this line of thought:

> Suppose we have a very simple physical system—a board in which there are two holes, a circle one inch in diameter and a square one inch high, and a cubical peg one-sixteenth of an inch less than one inch high. We have the following very simple fact to explain: *the peg passes through the square hole, and it does not pass through*

[14] See, for example, Fodor (1974), Putnam (1975: ch. 14), and Dennett (1987: ch. 2).

the round hole... Now, one can say that in this explanation [one citing geometrical properties] certain *relevant structural features of the situation* are brought out. The geometrical facts are brought out. It is *relevant* that a square one inch high is bigger than a circle one inch around... And nothing else is relevant. The same explanation will go in any world (whatever the microstructure) in which those *higher level structural features* are present. In that sense *this explanation is autonomous.*[15]

I take it that Putnam and others who emphasize explanatory autonomy are primarily arguing that the explanations offered by higher-level sciences that range over multiply realized kinds are *distinct* from lower-level explanations and are *preferable* in certain explanatory contexts. The claim of explanatory distinctness means that new insight, beyond that offered by the totality of lower-level explanations, can be gained by attending to the higher-level explanation. The relevant contextual factors that contribute to determining the preferability of an explanation include our interests and our cognitive limitations—e.g., we do not care about the very precise manner in which the peg passes through the square hole nor are we capable of keeping all of the microstructural details in mind. Among such theorists, I think that arguments of the following form would be widely accepted:

MR Argument for Explanatory Autonomy

EP1. Whenever some kind *K* can be realized in various *S*-kinds or collections of *S*-kinds, S_1–S_n (where *S* is a science or conceptual scheme), the explanations, predictions, and generalizations that essentially invoke only predicates for *K* are distinct from the explanations, predictions, and generalizations that essentially invoke only predicates for *S*-kinds or collections of *S*-kinds, S_1–S_n. In such cases the *K*-kind explanations, predictions, and generalizations are not merely abbreviations or notational variants of the *S*-kind explanations, predictions, and generalizations. Rather, the *K*-kind explanations, predictions, and generalizations are explanatorily autonomous with respect to the *S*-kind explanations, predictions, and generalizations.

EP2. Kind *A* can be realized in various *S*-kinds or collections of *S*-kinds, S_1–S_n.

EC. Therefore, the *A*-kind explanations, predictions, and generalizations are not merely abbreviations or notational variants of the *S*-kind explanations, predictions, and generalizations. Rather, the

[15] Putnam (1975: 295–296), italics in original.

> *A*-kind explanations, predictions, and generalizations are explanatorily autonomous with respect to the *S*-kind explanations, predictions, and generalizations.

Again, I allege that it is a virtue of an account of MR that it explains why MR is connected to explanatory autonomy in the way offered by EP1.

As with ontological autonomy, a distinction can also be made between weaker and stronger versions of explanatory autonomy. A weaker version is the claim of simple explanatory distinctness and preferability, as provided above. A stronger claim is that higher-level explanations that essentially invoke predicates for multiply realized kinds exist for some phenomena that lower-level sciences cannot explain. That is, for at least some *explananda* the only possible explanations are in terms of multiply realized kinds. On this stronger version, microphysics, say, cannot explain, even poorly or inappropriately, why the peg passes through the square hole.[16] The weaker version allows for such explanations, but simply asserts that the higher-level explanation is distinct from and, in some situations at least, preferable to lower-level explanations.

4.3.3 *Methodological Autonomy*

If multiply realized kinds are ontologically autonomous, then it also might not come as a surprise if they are, in some sense, methodologically autonomous as well. For if a multiply realized kind has a wide variety of realizations, why think that the methods that are appropriate for investigating these realizations are appropriate for investigating the multiply realized kind? They obviously operate at different levels of abstraction and, if explanatory autonomy is also correct, are invoked in distinct explanations, so it seems that the methodologies will likely differ as well.

MR Argument for Methodological Autonomy

> MP1. Whenever some kind *K* can be realized in various *S*-kinds or collections of *S*-kinds, S_1–S_n (where *S* is a science or conceptual scheme), productive inquiry into the explanations, predictions, and generalizations invoking predicates for kind *K* does not require adopting the methodology distinctive of *S*-kind inquiry. In such cases the science

[16] See Sober (1999) for criticism of this claim.

or conceptual scheme of the *K*-kind does not methodologically reduce to the *S*-kind science or conceptual scheme, and the *K*-kind science or conceptual scheme is methodologically autonomous with respect to the *S*-kind science or conceptual scheme.

MP2. Kind *A* can be realized in various *S*-kinds or collections of *S*-kinds, S_1–S_n.

MC. Therefore, productive inquiry into the explanations, predictions, and generalizations invoking predicates for kind *A* does not require adopting the methodology distinctive of *S*-kind inquiry. The science or conceptual scheme of the *A*-kind does not methodologically reduce to the *S*-kind science or conceptual scheme, and the *A*-kind science or conceptual scheme is methodologically autonomous with respect to the *S*-kind science or conceptual scheme.

Once again, I take it to be a virtue of an account of MR that it explains why MR is connected to methodological autonomy in the way offered by MP1.

Here I think that it is helpful to return to the origins of realization and functionalism in the philosophy of mind. For each, the computer metaphor was dominant. Recall that realization was originally introduced as that relation between abstract computational or functional states and the hardware that implements them. Such states are said to be multiply realizable in that various hardware could implement these same functions. According to functionalism in the philosophy of mind, mental states are functional states. This suggests that they too can be realized in various hardware. The methodological autonomy of psychology can then be justified on analogy with computers. One can be a productive programmer of a high-level computer language while remaining ignorant of machine or assembly language and the approach to computing taken from those levels. Likewise, one can be a productive psychologist while remaining ignorant of neuroscience and the approach to mentality taken from that level.

As with our other varieties of autonomy, there is a meaningful distinction between weaker and stronger versions of methodological autonomy. Unfortunately, many have easily slipped from endorsing the more plausible weaker version to the suspect stronger version. On the stronger version of methodological autonomy, adopting the methodology of the lower-level sciences cannot aid in the development of a science of multiply realized kinds. Whereas the stronger version states that it would not even help higher-level inquiry, the weaker version merely states that knowledge of the lower-level science is not necessary for productive higher-level inquiry (though it might be necessary for comprehensive or maximal higher-level productivity). We know that the stronger version is implausible in the computer case—knowledge of machine or assembly language likely *will* make one a better computer programmer with respect to high-level languages. Continuing the comparison, we should also expect that knowledge of, say, neuroscience likely will make one a better psychologist, even if psychology is weakly methodologically autonomous with respect to neuroscience.

4.4 Objections to Alternative Accounts

Before detailing my positive account of MR, which explains the role MR plays in these three arguments, I want to contrast my position with those of some prominent theorists. My most general and common criticism is that they fail to heed the distinction between determination and realization. As a result, their conception of MR renders it incapable of doing the desirable theoretical work to which it has traditionally been employed.

4.4.1 *Confusing Realization with Determination*

Stephen Yablo was one of the first philosophers to invoke the determinable–determinate relation in an attempt to account for mental–physical relations. In Yablo (1992) he argued that the realization relation that holds between mental and physical kinds is a species of the determination relation.

> . . . mental/physical relations are a species of determinable/determinate relations. "Can you really be saying that mental properties stand to their physical realizations in the relation that rectangularity bears to squareness, or that colors bear to their shades?" Yes. At least that is my conjecture. . . [17]

Some years later Sydney Shoemaker made a similar claim, asserting that determinate properties *realize* determinable properties.

> Being red, together with being of appropriate size, confers the power of evoking a pecking response from a pigeon such as Sophie under certain conditions. And so does the property of being scarlet. The latter, of course, confers powers not conferred by redness. For example, it confers the power of evoking a pecking response

[17] Yablo (1992: 256).

> in the likes of Sophie's sister Alice, who was conditioned to peck at scarlet things but not at things of other shades of red. In such a case we could say that the determinate property realizes the determinable in virtue of the fact that the conditional powers bestowed by the latter are a proper subset of those bestowed by the former. . . Where the realized property is multiply realizable, the conditional powers bestowed by it will be a proper subset of the sets bestowed by each of the realizer properties.[18]

Shoemaker later backed off this claim somewhat. He now holds that determination is a species of realization, and not every instance of realization is an instance of determination.[19] As plausible as these claims might be at first glance, upon further inspection they lead to some absurd conclusions.[20]

We should distinguish the specific claims that Yablo and Shoemaker are making when they each claim that realization is, at least in the mental–physical case, determination. Yablo suggests that the realization relation that holds between physical and mental kinds is the very same relation as that which holds between the kinds *square* and *rectangular, scarlet* and *red*. Shoemaker explains determination as inheritance of a proper subset of the causal powers of a kind—namely, the conditional powers of determinable kinds are proper subsets of the conditional powers of their determinate kinds. He then claims that the same kind of causal power inheritance is true of realization. I believe that both of these claims from Yablo and Shoemaker are false.

If the determination relation were a species of the realization relation, then MR would obtain whenever a determinable kind has multiple determinates. This seems to follow from the last two sentences quoted from Shoemaker, above. But MR does *not* obtain simply because a determinable possesses multiple determinates. For example, *red* is not multiply realized simply because it comes in various, determinate shades—such as *scarlet* and *crimson*. If that were the case, then *pain* would likewise be multiply realized simply because it comes in various, determinate kinds—such as *sharp pain* and *dull pain*. It should be obvious that this is not sufficient for MR, however. Otherwise, Putnam would have had no need to talk about pain in octopuses and reptiles to establish the MR thesis for psychology.

[18] Shoemaker (2001: 78–79). [19] Shoemaker (2007: 23).

[20] Examples of such confusion are also found in Clapp (2001: 125) and Ehring (2003: 375).

Nor is inheritance of causal powers sufficient to establish realization. I assume that the causal powers of, say, *red*, are a proper subset of the causal powers of *scarlet*. Likewise, the causal powers of *red* are a proper subset of the causal powers of *crimson*. But we have just seen that neither *scarlet* and *red* nor *crimson* and *red* are related by realization. Rather, they are related by determination. Again, this is most obvious when we remember that *red* is not multiply realized simply in virtue of coming in determinate shades *scarlet* and *crimson*.

One might retort that I simply have a different concept of MR in mind than do Yablo and Shoemaker. Perhaps. But we should ask which concept does the interesting theoretical work with respect to mind–body relations and in other scientific domains. This is a metaphilosophical principle that should have wide acceptance. It is easy for philosophers to make distinctions, but we should do so in a manner that yields interesting and fruitful work (theoretical or practical). In the previous section I argued that the interesting theoretical work that MR is supposed to perform lies with its connections to various concepts of non-reductionism and autonomy. This claim is certainly justified if one examines its genesis, as evidenced in the works of Putnam and Fodor. But, more importantly, I do not see that any interesting conclusions concerning the autonomy of multiply realized kinds follow on the Yablo–Shoemaker understanding of MR. Again, if the determination relation is taken to be a species of the realization relation, then *red* turns out to be multiply realized simply because it has various shades. But it would be absurd to conclude that *red* is thereby ontologically, explanatorily, or methodologically autonomous with respect to its shades. For example, no one should conclude that inquiry into *red* is autonomous with respect to inquiry into its various shades! There is a sharp distinction between determination and realization—in fact, they are mutually exclusive relations. This understanding of MR better fits the actual use to which the concept has been fruitfully employed.

4.4.2 *Different Ways of Performing a Function*

We saw that Yablo and Shoemaker explicitly conflate, in at least some instances, determination and realization. Others are guilty of doing so implicitly. Ruth Millikan and Lawrence Shapiro, who argue that multiple realizations are different *ways* of performing the same function, belong in this category. I will argue that their different "ways" are in fact different determinates of a common determinable and not multiple realizations of a common kind.

It was once widely thought, if not still, that functional kinds must be multiply realizable and that functionalism is therefore manifestly a non-reductive philosophy of mind. But Shapiro (2000, 2004) argues that only some functional kinds are multiply realizable. In contrast with the largely *a priori* nature of many of the early functionalists' arguments for MR, Shapiro claims that MR requires that there be multiple ways of performing the relevant function. And, as he uses 'way,' not every difference in a realization counts as a different way of performing that function.[21] Shapiro emphasizes that it requires empirical work to determine if a given function can be performed in multiple ways. He explains his proposal as follows, via the example of corkscrews:

> Some of the properties of the realizers of multiply realizable kinds are relevant to the purpose, activity, or capacity that define the kind and some are not. I asked in the introduction whether two corkscrews, alike in constitution and mechanism but distinct in color, count as alternative realizations of the kind corkscrew. I claim that they do not because the only property by which they differ—color—is not a property that contributes to their capacity to remove corks. To say that a kind is multiply realizable is to say that there are different ways to bring about the function that defines the kind.[22]

Shapiro's key move is to distinguish those properties that are relevant to fulfilling the constitutive function of a functional kind from all the other realization properties. Only differences in the former properties amount to differences of realization. This division of properties corresponds to one that we have already given emphasis. Those properties that are relevant to fulfilling the function that is constitutive of a functional kind simply are those at that level of abstraction—i.e., those providing the determination dimensions of that functional kind. Functional kinds, on Shapiro's usage, are abstractions that isolate certain capacities which contribute to

[21] Shapiro (2000, 2004) can be seen as a development of Millikan's (1999) passing comments: "Sometimes different mechanisms that accomplish the same operate [sic] in accordance with different principles; other times they represent merely different embodiments of the same principles. Or we might say, sometimes looking more closely at the mechanism helps to explain how it works; sometimes it reveals only what stuff it is made of. It is only the former kind of difference that makes interesting 'multiple realizability'" (61–62). As Shapiro (2000, 2004) offers a more developed presentation of this line of thought, I will limit my focus to his discussion.

[22] Shapiro (2000: 643–644).

performing some function. Given that these capacities can vary but are central to the essence of functional kinds, these capacities seem to provide us with the determination dimensions for their kinds. So, variations along these dimensions represent different determinates of that functional kind rather than different realizations. Features that are not relevant to the performance of that function are not features in which instances of a functional kind can differ *qua* that functional kind. For example, instances of *corkscrew*, understood as a functional kind with the function of removing corks, can differ *qua* corkscrew[23] with respect to the mechanism by which they remove corks, but not with respect to their color or material constitution. Compare: colors can differ with respect to hue, brightness, or saturation, but not with respect to chemical makeup, etc. Also recall the example of *f-pain*, as discussed in Chapter 2, for a more scientific example of a functional kind and its determination dimensions.

Mapping Shapiro's division of relevant and non-relevant realization properties onto our division between multiple determinates and multiple realizations of a kind helps to reveal what is mistaken about his restrictions on MR. Shapiro's different ways of fulfilling a function are different determinates of the same determinable. The different mechanisms of corkscrews are different determinates along the abstraction 'cork-removing techniques.' Whereas Shapiro judges the "waiter's" and "winged" varieties to be multiple realizations of *corkscrew*, our theory judges them to be multiple determinates of the kind *corkscrew*. The science of *corkscrews*, if there were such a thing, should mark this distinction between "waiter's" and "winged," as this very distinction is relevant to individuating corkscrews. Of course, I do not think that this kind should be taken with much seriousness. Other examples are more appropriate here. The science of *p-pain* should mark the distinction between *sharp pain* and *dull pain*, and the science of *memory* should mark the distinction between *short-term memory* and *long-term memory*.

Again, the basic idea of MR is that there is sameness of *A*-kind through differences in *S*-kind, though the *A*-kind is realized by the *S*-kinds. The mental is multiply realizable in the physical if and only if there can be sameness of super-determinate mental kind through differences in the

[23] One problem with the corkscrew example is that *corkscrew* is a substantival kind, and corkscrews are objects rather than properties. To treat them as abstractions, I have resorted to the '*qua*' locution.

physical kinds, even though the mental kind is realizable by those physical kinds. Now let us apply this to the corkscrew example. *Corkscrew* MR consists in sameness of *corkscrew* kind through differences in some *S*-kinds, though the *corkscrew* kind is realizable by the *S*-kinds. Shapiro offers the different mechanisms—"waiter's" and "winged"—as different *S*-kinds that serve as multiple realizations. But note that a difference in corkscrew mechanism is necessarily a "corkscrew difference." Indeed, this seems to be Shapiro's whole point. But, importantly, *multiple realizations of an A-kind do not necessarily make an A-kind difference*! Multiple realizations of a mental kind like *pain*, if they exist, do not necessarily differ in their "pain-ness." In contrast, corkscrews operating according to different mechanisms thereby necessarily differ in their "corkscrew-ness."

4.4.3 Disjunctions, Heterogeneity, and Kim's Challenge

In §4.3 I presented three simple arguments for non-reductionist conclusions from premises concerning MR. I accept the weaker versions of these schematic conclusions for each sense of autonomy for the various special sciences and their kinds. But while these conclusions are widely accepted by philosophers of mind and cognitive scientists, there have been some prominent dissenters. Notable among these is Jaegwon Kim, who has denied that MR entails ontological, explanatory, or methodological non-reductionism. Kim (1993a), in particular, offers a sustained attack against premises connecting MR to autonomy. In fact, he argues for the opposite conclusion—MR entails local reductions in ontology, explanation, and methodology.

Let us consider his objection driven by "the disjunctive move." All parties can agree that if a kind is multiply realizable, then its possible realizations are heterogeneous. Kim takes *pain* as a candidate for a multiply realized kind. To simplify matters, he harmlessly stipulates that it has three heterogeneous neuroscientific realizations: N_h, N_r, and N_m (corresponding to *human pain, reptile pain*, and *Martian pain*). So, the kind *pain* is nomically equivalent to the disjunction $N_h v N_r v N_m$. Fodor has argued that such heterogenous disjunctions of neuroscientific kinds are not themselves neuroscientific *kinds*. As such, terms for these disjunctions cannot legitimately appear in scientific laws and generalizations. In particular, Fodor has argued that terms for such disjunctions of neuroscientific kinds cannot appear in the bridge laws needed to secure theoretical and ontological reductions. So, they cannot serve as candidates for the reduction of *pain*.[24]

But Kim counters Fodor as follows: since *pain* is nomically equivalent to $N_h vN_r vN_m$, pain predicates pick out kinds that are just as heterogeneous and illegitimate for scientific laws and generalizations as do predicates for the corresponding "disjunctive kinds." Kim asks the rhetorical question:

> If pain is nomically equivalent to *N*, the property claimed to be wildly disjunctive and obviously nonnomic, *why isn't pain itself equally heterogenous and nonnomic as a kind?*[25]

Kim concludes that both kinds, *pain* and $N_h vN_r vN_m$, are on the same footing in this regard. The consequence is supposed to be that, due to their heterogeneity, there cannot be a productive science of multiply realizable kinds. In effect, we can see Kim as issuing the following challenge:

> **Kim's Challenge:** Provide the interesting (e.g., non-analytic) laws, generalizations, or explanations that any science has discovered concerning multiply realized kinds.

Kim, of course, thinks that this challenge cannot be met. Instead, the sciences that purport to study these kinds should fragment into distinct sciences corresponding to each realization kind. Rather than psychology *simpliciter*, there should be a division between human psychology, reptile psychology, and Martian psychology. Kim insists that if *pain* is multiply realized, then we should not expect generalizations about *pain* as confirmed in human cases to hold for reptiles or Martians. Rather, according to Kim, MR should lead to conceptual fragmentation (i.e., from pain to human pain, reptile pain, and Martian pain) and then to corresponding ontological, explanatory, and methodological reductions.[26]

The main problem with Kim's argument is that though *pain* is nomically equivalent to $N_h vN_r vN_m$, they are not equally heterogeneous. No doubt the *creatures* that can be of these kinds are equally heterogeneous, as the set of creatures that can be in pain is, by stipulation, one and the same as the set of creatures that can be in neuroscientific states N_h, N_r, or N_m. But this does

[24] Fodor (1974). In his discussion Fodor is using 'kind' in a more restrictive sense than we are using it. By 'kind' he means something like 'natural kind.'

[25] Kim (1993b: 323).

[26] For responses to Kim's argument, see Fodor (1997) and Block (1997).

not show that the kinds themselves are equally heterogeneous. By stipulation, N_h, N_r, and N_m differ wildly with respect to the classifications of neuroscience. It is helpful to apply our theory from Chapter 2 at this point. We can now see that N_h, N_r, and N_m are neuroscientific kinds with distinct absolute non-determinable necessities, distinct determination dimensions, or wildly different determination dimension values. In no other sense could they be wildly heterogeneous. Further, given the claim of MR we know that *pain* and $N_h vN_r vN_m$ do not share determination dimensions. The determination dimensions for the psychological kind *pain* are likely quite different than the determination dimensions for N_h, N_r, or N_m (or their disjunction). For example, pains might be individuated, at least in part, on phenomenological grounds, whereas N_h, N_r, and N_m are not individuated on such grounds. The moral is that all *pain* kinds share the same determination dimensions, though their realization kinds possess different determination dimensions. Further, these realization kinds might not, and often do not, share determination dimensions among themselves.

We can also object to Kim by disproving the conclusion he draws from this premise—localized reductions and skepticism about sciences ranging over multiply realized kinds. We should always be skeptical when a philosopher argues on *a priori* or metaphysical grounds for the conclusion that some apparent scientific possibility is in fact impossible. This is especially true when practicing, well-funded scientists are engaged in the metaphysically prohibited activity. And here we find Kim making a claim, based on metaphysical considerations alone, that can be empirically falsified. I have no problem with testable metaphysical claims, but I think that Kim is mistaken in his claim that there cannot be a science of multiply realized kinds as such. To refute this claim, we need only provide examples in which special science generalizations hold across objects or systems that differ wildly in their realizations. I do not share Kim's conviction that such examples are not to be found.

Other metaphysicians and philosophers of science—such as Ruth Millikan (1999) and Richard Boyd (1999)—have argued that multiply realized kinds, insofar as they are functional kinds, cannot serve as the predicates of an empirical science. Millikan states her conclusion as follows:

> Indeed, I will argue, the idea that there could be a univocal empirical science that ranged over multiple realizations of a functional property is quite problematic. For example, if psychological predicates name multiply realized functionally defined properties, then there can be no single science that deals with all items having

> these properties, human psychology, ape psychology, Martian psychology and robot psychology are necessarily different sciences.[27]

There is a shared rationale supporting the conclusion of Kim, Millikan, and Boyd. Basically, the idea is that the multiple realizations of a functional kind would be so diverse that we should not expect them to fall under any common regularities, but for those given in their functional specification.[28]

This position obviously counters the optimism of those trumpeting the explanatory and methodological autonomy of psychology. But how does the present account of realization and MR affect this debate? It should be noted that Millikan and Boyd (and perhaps even Kim) are directing their arguments at multiply realizable *functional* kinds. But on the present account of realization non-functional kinds can have realizations (and multiple realizations) as well. In one of our examples it was shown that even if *pain* were a strictly non-functional kind (i.e., *p-pain*), it could still be multiply realizable. It is easy to see how this point generalizes so that many other non-functional kinds could be multiply realized. The insight is that MR goes hand in hand with operating at a different "level of abstraction." So even if Millikan and Boyd are correct that there cannot be an empirical science ranging over functional kinds, they have not shown that there cannot be an empirical science ranging over non-functional, multiply realized kinds. Alternatively, if their argument is meant to equally apply to non-functional, multiply realized kinds, then it is clearly flawed as there manifestly are empirical sciences concerning such kinds.

Presenting a full response to Kim's Challenge and defending the legitimacy of sciences of multiply realized kinds is beyond the scope of this book. (Some particular examples are discussed in §5.3, however.) However, I feel obliged to state my sympathy for the view that there can be, and in fact are, empirical sciences and generalizations concerning functional kinds. It seems clear that there are empirical sciences ranging over non-functional multiply realized kinds. And why shouldn't we expect the same for multiply realized functional kinds? We can object to Kim, Millikan, and Boyd by pointing out that the reasoning they use against an

[27] Millikan (1999: 47).

[28] Kim (1993a: 324–327); Millikan (1999: 50, 54–55, and 58–60); Boyd (1999: 69–72).

empirical science of functional kinds equally holds against sciences ranging over non-functional multiply realized kinds. But surely there are sciences of the latter type. If, instead, all such supposed examples are in fact examples of multiply realized functional kinds, then the response is even more direct and effective. For example, chemistry invokes predicates that pick out the kind *acidic*. But we could apply their same rationale against even this possibility: just as radically different creatures can be of the same psychological kind in virtue of radically different realization bases, radically different compounds can be of the same *acidic* kind in virtue of radically different realization bases. We should not expect there to be any interesting (i.e., *a posteriori*) generalizations that hold across radically different realizations of the same psychological (functional) kind. Similarly, we should not expect there to be any interesting (i.e., *a posteriori*) generalizations that hold across radically different realizations of the same *acidic* kinds. And this form of argument seems wholly indifferent to whether or not the essence of the multiply realized kind is functional.

Without being able to explain *how* there can be a science of multiply realized kinds, I simply believe that there are such sciences. They are all around us. Fodor, I believe, gives a good response to Kim's Challenge by making, to my mind, this very obvious point.

> The very *existence* of the special sciences testifies to reliable macrolevel regularities that are realized by mechanisms whose physical substance is quite typically heterogeneous. Does anybody really doubt that mountains are made of all sorts of stuff? Does anybody really think that, since they are, generalizations about mountains-as-such won't continue to serve geology in good stead? Damn near everything we know about the world suggests that unimaginably complicated to-ings and fro-ings of bits and pieces at the extreme *micro*level manage somehow to converge on stable *macro*level properties.[29]

We can leave it to those who study complex systems theory to attempt to explain how, in general, higher-level order emerges from lower-level heterogeneity. Or we can leave it to the practitioners of a particular science to explain this phenomenon for a particular domain. And I am more than comfortable conceding that the armchair metaphysician is ill-prepared for these tasks.

[29] Fodor (1997: 160).

4.5 The Mutual Exclusivity of Determination and Realization (Again)

One of the strongest and broadest claims I have advanced is that determination and realization are mutually exclusive relations. We saw arguments for this in the concluding paragraphs of both §4.1 and §4.2. But given the importance of this point, I wish to close this chapter by returning to the main reasons for this claim. In clearly distinguishing realization from determination, my view of realization contrasts with that of others (e.g., Yablo, Shoemaker, and Shapiro). The claim might also be somewhat startling in its own right. Using our property theory and our understanding of realization, let's review why these relations turn out to be mutually exclusive.

First, why can't realizers determine their realizations? If realizing kinds could determine their realization kinds, then realizer kinds could be a more specific way of being that realization kind. On our property theory, kinds and the properties that are their instances are abstractions whose essences are constituted by determination dimensions (for kinds), determination dimension values (for properties), and non-determinable necessities (for both). But realizer kinds possess different determination dimensions than do their realization kinds. This is what allows for the possibility of MR—sameness of one set of determination dimension values (e.g., psychological values) through differences in another set of determination dimension values (e.g., neuroscientific values).[30] But if realizer and realization kinds have different determination dimensions, then the former cannot provide more specific determination dimension values. Realizer kinds can, of course, necessitate specific realization values, as neuroscientific kinds can necessitate specific psychological kinds. But these neuroscientific kinds, like all kinds, are abstractions and do not themselves have those determination dimensions. And kinds can be related as determinable to determinate only if they share determination dimensions (i.e., are at the same level of abstraction).

The main line of argument that I have advanced against realization being a species of determination—based on considerations of MR as just discussed—is very similar to one found in Ehring (1996).[31] But Ehring

[30] As I argue in §5.5.3, *all* realization is multiple realization. So-called singular realizability is best conceived as identity.

[31] See his Objection 3, in Ehring (1996: 473).

merely argued that the physical does not determine the mental, not that the relations are mutually exclusive. Ehring provided four arguments for the claim that the physical does not determine, but instead only realizes, the mental. I accept at least the latter three of these arguments,[32] and the property theory I have offered provides a deeper explanation of these points and generalizes them well beyond the case of the mental–physical. Ehring properly notes that physical realizers sometimes have an ordering that diverges from the ordering of the mental properties they realize. This is especially true given functionalism and MR. Quite similar mental kinds (or even the very same mental kind) can be realized by quite different physical kinds. There may be no neat ordering of these physical realizers. But determinates of a common determinable thereby share a similarity and ordering (e.g., *orange* is in between *red* and *yellow*). Further, he notes that the distinct physical realizers of a common mental determinable need not differ in their mentality. But determinates of a determinable always differ in that regard.[33] These points should now sound familiar. The reason why these realizations do not possess the similarities (orderings) and differences that determination requires is because realization kinds possess distinct determination dimensions from those kinds that they realize. Since these kinds and properties just are abstractions along these dimensions (in conjunction with their non-determinable necessities), it is impossible for the realization and realized properties to be similar or dissimilar in these regards.

Next, why can't determinates realize their determinables? First, we can see that this is a consequence of the present understanding of realization and determination. Determinates and their determinables share determination dimensions, but realized kinds and their realizations do not. In this sense, they are at different levels of abstraction. Determinates of some kind *K* necessarily differ with respect to their *K*-ness. But realizations of some kind *K* do not necessarily differ in their *K*-ness. In fact, and in stark contrast with the determinates of *K*, the realizations of *K* are *completely*

[32] Ehring's first argument alleges that there can be more than one same-level physical realizer, but only one same-level determinate. I am not sure that I accept this consequence, as such physical realizers are not necessarily instantiated at the same spatial-temporal location or by the same object. Such a case might be more similar to an object having different super-determinate colors at different parts.

[33] Ehring (1996: 473).

blind to *K*-ness. Possessing entirely distinct determination dimensions, they abstract away from the *K*-relevant content. Because realization and realized kinds possess distinct determination dimensions, the properties which are their instances are always distinct. But determinable kinds, in contrast, share instances with their determinate kinds. An instance of *red*, for example, is identical to an instance of *scarlet*. If this were also an example of realization—with the *scarlet* property realizing the *red* property—then they would be distinct properties. The present theory denies this, however. And this should be a welcome consequence. The alternative view would proliferate qualitatively indistinguishable properties that realize one another—e.g., a *scarlet* property that realizes a *red* property that realizes a *color* property.

But why is this a desirable, or at least acceptable, consequence? Why demand this of realization? I suppose that one could understand realization as a generic necessitation relation that covers both determination—so that the kind *scarlet* realizes *red* (even though they share instances)—and my more restrictive understanding of realization. So our second reason is that the widely accepted connection between realization and autonomy claims, as discussed in §4.3, tells against this. It is possible to have a fruitful investigation of a realized kind while remaining ignorant of realizer details *because* they possess different determination dimensions (i.e., operate at different levels of abstraction). But you cannot fruitfully study *red* without studying *scarlet, crimson*, and so on. In some limited sense, you can study a determinable kind without studying its determinates. You can discover true generalizations, say, about *red* while remaining in the dark about its determinate shades. But one cannot be much of an expert about *red* without knowing about its determinates. Try to imagine such a situation. It would be like someone claiming to be a wine expert, but being oblivious to the distinctions between Pinot Noir, Merlot, etc. Here is a helpful contrast: an expert about *red* must be knowledgeable about its determinate shades; an expert at cognitive psychology need not be knowledgeable of its neuroscientific or chemical realizations (assuming the MR of the psychological kinds, and the autonomy that comes with it).

Another way of putting this point is that realizers are different in kind from their realizations, not merely different in degree (specificity). This leads us to our third reason for denying that determinates realize their determinables. Determinates are not different in kind from their determinables, in the sense that they share determination dimensions and differ only in their specificity along those dimensions. Realizations, in contrast, are different in kind from that which they realize. This is what leads us to claim property distinctness for realizations and that which they realize. It would be helpful to have distinct relations to account for this common-sensical distinction between a difference of kind (realization) and a mere difference of degree (determination). And differences of kind and differences of degree are mutually exclusive.

5

Multiple Realizability II: An Analysis

I have argued that realization should be carefully distinguished from determination and that MR is the critical relation in various arguments for non-reductionism and autonomy in the special sciences. It is now time to offer a positive account of MR that meets these requirements.

5.1 An Analysis

The contrast with multiple determination jump-starts the present analysis of MR. I suggest the following extremely simple conditions for MR that capture our strengthened idea of "sameness through difference":

Necessary and Sufficient Conditions for Multiple Realizability

Kind *A* is multiply realizable if and only if there are nomologically possible instances of *A* that:

1. are of the same super-determinate or maximally specific *A*-kind (e.g., share the same precise determination dimension values and non-determinable necessities), though;
2. the realization bases for these *A*-properties differ with respect to some set of determination or specification dimension values.

Here the realization bases referred to in condition 2 should be understood according to the accounts of total and core realization offered in §4.2. These realization bases could be restricted to core realizations, or they could be total realizations. This distinction gives rise to two other senses of MR—core MR and total MR. Again, given the interest in scientific kinds,

I imagine that core MR will be the more interesting and beneficial concept. It should then be clear what is required for *pain*, say, to be multiply realizable according to this analysis. *Pain* is multiply realizable if and only if it is possible to have exact pain resemblance through differences in the lower-level properties in which those pains are realized.[1]

Note that the modality involved here is nomological necessity rather than, say, metaphysical necessity. Realization itself, however, is still taken to be a relation of metaphysical necessity. For example, brains realize consciousness only if they do so with metaphysical necessity. But I think it is beneficial to limit claims of MR to those kinds whose multiple realizations are at least nomologically possible. This is because MR claims are most common and interesting for scientific kinds. The presence of merely metaphysically possible realizations—like soul-stuff (where such stuff is assumed to be metaphysically possible, but not nomologically possible) realizing consciousness—is not enough to secure the explanatory or methodological autonomy of psychology in the actual world. Such remote possibilities are irrelevant to actual world psychological explanation and methodology. They very likely are relevant to ontological claims, however. So, we might want to employ different notions of MR—nomological or metaphysical—in different contexts. I will typically employ the concept in its nomological sense, and this is perfectly consistent with realization itself being a relation of metaphysical necessity. But to be clear, I also tolerate a looser, metaphysical sense of MR.

This analysis has two necessary conditions that are jointly sufficient for MR. Each condition is novel and bold. Condition 1 provides a restriction on MR, disqualifying mere cases of multiple determination as cases of multiple realization. We have already given a sustained defense and articulation of this distinction and this claim. This emphasis is justified, as we have just observed that many have failed to mark or appreciate the significance of this distinction—e.g., Yablo, Shoemaker, Millikan, Shapiro, and Kim. Many will probably be drawn to the bold liberality of condition 2. Though ruling out cases of kinds related as determinable to determinate, the present analysis otherwise quite readily countenances MR. This liberality will be defended in the next section.

[1] Polger (2002: 152) also requires such exact resemblances for the type of MR that is supposed to pose a problem for kind identity theories.

A slightly more liberal proposal would be to countenance MR whenever there is very close pain resemblance through relatively significant differences in the lower-level properties in which those pains are realized. After all, in most actual situations we can only have confidence that there are close, as opposed to exact, psychological resemblances. This is an epistemic point, however. But if close pain resemblances can obtain through radical lower-level differences, this gives us reason to think that exact pain resemblance *could* obtain through diverse lower-level differences as well. And the present account of MR only requires this nomological possibility. So as a methodological point, we should take near sameness of psychological kind through lower-level heterogeneity as a good guide to MR.

It should be emphasized that the same kind can be both multiply determined and multiply realized, though such a kind would have to bear these relations to distinct kinds. That is, no kind *A* can bear *both* the determination and realization relation to any single kind *B*. Though, it might be that kind *A* is determined by kind *B* and realized by some other kind *C*. Indeed, it is very plausible to think that *pain* admits of both varieties of qualification. *Sharp pain* and *dull pain* are two determinates of *pain*, since they differ in their "pain-ness." So *pain* has multiple determinates. And each instance of *pain* is of some super-determinate kind, just as each instance of *color* is some very particular shade. But if realizations of the same super-determinate *pain* kind can differ at some lower level of abstraction, then it follows from our analysis that this super-determinate *pain* kind, and *pain simpliciter*, is multiply realizable.

John Heil has cautioned that when developing accounts of MR we should be "ontologically serious" about the key notions—e.g., realization and property.[2] I think that I have been quite ontologically serious thus far, but I do not think that such seriousness is necessary to adequately account for or explain MR. In fact, the present account of MR can be characterized without any ontological seriousness: some classification is multiply realizable just in case a completely accurate and specific classification of that kind need not pay heed to at least some of the mechanistic details in virtue of which objects belong to that classification, where these mechanistic details can vary though the classification does not vary (even in its specification). And this characterization rings true whether or not we think that the properties and kinds corresponding to these classifications *really* exist.

[2] Heil (2005).

5.2 Objections and Replies

Objection 1. Super-determinate sameness through realization differences is not necessary for MR.

Matthew Haug has argued against my claim that MR obtains only when a super-determinate kind has multiple realizations. In an article otherwise largely sympathetic to the present account, Haug voices the following dissent:

> Clearly, a sufficient condition for a determinable to be multiply realized is that one of its maximally determinate properties is multiply realized. However, this is not a necessary condition on the multiple realization of high-level determinable properties. Suppose that there is only one possible realizer, R_1, of a maximally determinate shade of red, S_1; there is only one physical constitutive mechanism that can "produce" this shade. Suppose also that this is sufficient to warrant the identity, "$R_1 = S_1$." Suppose that the same is true of a distinct maximally determinate shade of red S_2; some *distinct kind* of mechanism uniquely realizes it, so $R_2 = S_2$. In this case, *being red* is multiply realized, but the distinct realizers *do differ* in their "redness," contra the passage from Funkhouser.[3]

The idea here is that a determinable kind, such as *red*, can be multiply realized in virtue of having different super-determinates, such as S_1 and S_2, each having singular realizations that are mechanistically distinct from one another. Haug's claim is that the MR of a determinable kind does not require the MR of any individual super-determinate of that kind, contrary to what I have claimed.

There are a few things to say about this supposed counterexample to the necessity claim. I have offered an account of the modal notion MR, so for Haug's counterexample to go through it must be the case that S_1 could not *possibly* be realized by an R_2 kind mechanism, and S_2 could not *possibly* be realized by an R_1 kind mechanism. I am not accusing Haug of overlooking this point. In fact, he actually does use the modal vocabulary. But I am not as confident that the situation he describes is as likely or clearly possible as he assumes. This would have to be a case in which *crimson*, say, is realized by a kind of mechanism the likes of which could not realize *scarlet*. There is also the issue of how to individuate his "kinds of mechanisms," because obviously no super-determinate realization kind is ever going to realize different super-determinates of the

[3] Haug (2010: 326).

same determinable. To the extent that we have an intuitive understanding of different "kinds of mechanisms," it is not clear that the situation he describes is possible.

But the more significant problem with the supposed counterexample is that it is effective only if *red* has *no* super-determinates with multiple possible realizations. Remember, the necessity claim at issue is that in order for a determinable kind to be multiply realizable, there must be some individual super-determinate of that kind with multiple possible realization kinds. Haug has only presented a case in which two of the super-determinates of *red* supposedly possess singular realizations. But my necessity condition is still met, and hence we do not have a counterexample, if there are other super-determinate shades of *red* that do have multiple possible realizations. And this is overwhelmingly likely. The situation Haug needs to serve as a counterexample is one in which every one of *red*'s super-determinates is singularly realizable, though *red* itself is supposedly multiply realizable. It is extremely unlikely that *red*, or almost any higher-level kind, would present such a situation. But let's suppose that it did.

Fully imagine how bizarre it would be for a determinable kind to have each of its super-determinates supposedly singularly realized, but by different kinds of mechanisms. In fact, we are to suppose that no single super-determinate is even multiply realiz*able*. So some super-determinate *scarlet* kinds can be realized only by $mechanism_1$ kinds, other super-determinate *scarlet* kinds can be realized only by $mechanism_2$ kinds, some super-determinate *crimson* kinds can be realized only by $mechanism_3$ kinds, and so on for each super-determinate variety of *red*. But these mechanism kinds are a heterogeneous lot. It would be odd, indeed, if things so similar as different shades of red could be realized in only one way each, but also in different ways distinctive to the different shades. If the realizers of the different shades of red were singular, we would expect similarities among them and either a reduction of *red* kinds to these realization kinds or an elimination of *red* kinds in favor of them. But such similarity is not supposed to hold. The objection to my view is that, in this unlikely (if not impossible) situation, the determinable kind *red* is in fact multiply realized, even though my theory does not have this result. But would this be an unsatisfactory result?

When we consider the connections between MR and autonomy, we find that this example presents an interesting and complicated situation.

When a kind is multiply realized, then it is autonomous—ontologically, explanatorily, and methodologically—with respect to its realizations. But there are senses in which the super-determinate kinds are not autonomous with respect to their realization kinds. Given that by necessity some super-determinate *crimson* instance occurs if *and only if* some *mechanism*$_3$ instance occurs as its realization, the method for studying the shades of crimson should be tied closely to the method for studying mechanism$_3$. The mark of a multiply realized kind, however, is insensitivity to realization detail, from which it follows that the methodology for studying the realized kind can ignore the methods for studying its various realization kinds. But this insensitivity to realization detail is lacking in the present example, given the assumption of its singular realizability (SR). Generally speaking, effective investigation of the colors would require studying the *unique* realization bases on which they depend. After all, each particular shade is supposed to derive from its own distinctive realization kind. Now, of course, the objector is not claiming that these super-determinate kinds are multiply realized; rather, the determinable kind *red* is claimed to be multiply realized. But every property is assumed to be an instance of some super-determinate kind (as well as an instance of the determinable kinds it falls under). And every super-determinate kind, by supposition in this example, has only one realization kind. Then, each *red* property has only one possible realization, and investigation of such *red* properties should then be more closely tethered to the investigation of its realization kinds than we would expect for cases of genuine MR.

Here is what is especially puzzling about this situation: the super-determinate *red* kinds are not methodologically autonomous with respect to their realization kinds. So to study the shades of red, we need to study their realization kinds. But it is still claimed that the determinable kind *red* is multiply realized. So to study the more determinable *color* kinds, we supposedly do not need to attend to realization-level detail. But determinables and their determinates are part of the very same conceptual scheme, operating at the same level of abstraction and differing only in specificity. We should expect that there is a shared methodology for studying determinables and their determinates, but that is not the case here. The methodological autonomy of the supposedly multiply realized determinable kind *red* is, at best, unclear.

Further, given that the various shades of red are uniquely, or at least distinctively, dependent on their singular realizations, an explanation citing a super-determinate *red* kind would not be explanatorily autonomous from an explanation citing its corresponding *mechanism*$_n$ kind. Whatever connection that super-determinate *red* kind bears to some *explanandum*, the *mechanism*$_n$ kind bears to it as well. This is especially true since there is a one-to-one correspondence between these kinds, and instances of the super-determinate kind always hold in virtue of instances of the realization kind. So the mechanism kinds will equally co-vary with these *explananda*, and—because of the "in virtue of" requirement on realization—they will also inherit an explanatory relationship to them.

Systematic SR, as I will argue in §5.5.3, suggests reduction, but if the realization mechanisms for the various super-determinate kinds are diverse enough then perhaps an elimination is in order. Haug appears to note the connection between such SR and identity (reduction) when he actually identifies the super-determinate shades, S_1 and S_2, with their singular realizations. However such an identification would not be possible on the present understanding of properties and realization, given the supposed difference in determination dimensions, were R_1 and R_2 truly realization kinds. S_1 and S_2 are supposed to share the traits characteristic of colors (e.g., hue, brightness, and saturation), whereas R_1 and R_2 are supposed to possess the different traits characteristic of their respective mechanistic kinds. But accepting the S_1=R_1 and S_2=R_2 identities entails that S_1 and S_2 belong to distinct property spaces (given the supposed distinctness of R_1 and R_2) rather than belonging to the common property space that we assume for all colors. Identifying the various shades of *red* with the mechanistic kinds that were supposed to be its realizations means, on the present view, that the property space of *red* is different than previously conceptualized and should be reduced, perhaps with a corrective fragmentation, to the property spaces for these mechanistic kinds. Alternatively, *red* would simply be eliminated in favor of these mechanism kinds.

Haug is correct that the simple discovery that different kinds of mechanisms realize different super-determinate kinds of the same, relatively well-established, determinable does provide us with some reason to think that the determinable kind is multiply realizable. I accept that his example has this force. But that is because it is then likely that an individual super-determinate of that determinable kind is multiply realizable.

To summarize, we have at least four responses to Haug's objection. First, we should note that it is unlikely that super-determinates of a common determinable would possess singular realizations distinct in kind from one another. Second, this is even more unlikely when this is claimed to hold with modal force, as is necessary to deny the multiple realiz*ability* of these super-determinate kinds. Third, in such an unusual situation we do not have the autonomy—especially the methodological autonomy—that we would expect of a genuine case of MR which is insensitive to lower-level detail. Since each super-determinate kind is necessarily dependent on a unique realization kind, it is much more likely that investigation of the super-determinate kind should attend to the details of its singular realization kind. This shows that the investigation of *red* is not so autonomous from the investigation of its realization kinds. Fourth, even if we grant autonomy to the determinable kind, we should also point out that we have only been arguing that MR entails various senses of autonomy. But we have not denied that there can be autonomy without MR. There may be rare cases in which a determinable kind like *red* is autonomous with respect to its singular realization kinds. But this is because we are supposing that nearby determinates have distinct realizations, which is an approximation of our demand for MR: sameness of super-determinate kind through realization differences. And a determinable is *more* multiply realized, and hence has greater autonomy with respect to its realization kinds, to the extent that more of its super-determinates are multiply realized.

Objection 2. But this analysis cheapens MR. Almost every kind is multiply realized in this sense!

If all that is required for MR is some realization difference across possible instances of the same super-determinate kind, no matter how fine or low-level this difference is, then multiply realizable kinds are ubiquitous. This might seem like an objection to the present conception of MR. Surely MR is somehow special and something is lost if so many kinds are multiply realizable. Worse yet, even standard examples of kinds that are not multiply realizable turn out to be multiply realizable on this conception. *Water* is a paradigm example of a substantival kind that is not multiply realizable, as its only metaphysically possible realization is widely thought to be H_2O. But some samples of the same maximally specific *water* kind surely differ at some extreme microphysical level.[4] Such a difference would be sufficient

[4] Perhaps *water* has no specifications, in which case we should simply speak of samples of *water* (*simpliciter*) that differ at some microphysical level.

to establish that even *water* is multiply realizable according to this analysis. Is this a refutation, a *reductio ad absurdum*, of the present conception? No.

This liberality is acceptable once we realize that there is also a concept of *relative* MR. Let us assume that samples of water can differ at some microphysical level of abstraction, while exactly resembling with respect to their "water-ness." Then *water* meets our conditions for MR *simpliciter*. This is because *water* is, by assumption, multiply realizable relative to microphysical kinds. But those who point out that water is not multiply realizable, with H_2O as its only possible realization, are still onto something correct. *Water* is not multiply realizable relative to that chemical level of abstraction. Further, the fact that *water* is realized by only one chemical kind explains why *water* is a chemical kind. If our assumption that *water* is multiply realizable in the microphysical is correct, then this also shows that *water* is not a microphysical kind. Our concept of relative MR *explains* these facts.[5]

The concept of relative MR should be understood as follows:

Necessary and Sufficient Conditions for Relative Multiple Realizability

> Kind *A* is multiply realizable relative to science or conceptual scheme *B* if and only if there are nomologically possible instances of *A* that:
>
> 1. are of the same super-determinate or maximally specific *A*-kind (e.g., share the same precise determination dimension values and non-determinable necessities), though
> 2. their *B* realization bases differ with respect to their *B* determination or specification dimension values.

Water is multiply realizable relative to microphysics because there can be sameness of maximally specific *water* kind through differences in microphysical determination or specification dimension values.

Objection 3. MR requires real heterogeneity, not just variability.

According to the present analysis, MR obtains whenever there are two or more realization kinds. But advocates of MR typically have not taken

[5] For the full version of this argument, see §5.5. Perhaps *water* could also be a chemical kind in virtue of being realized by a small family of closely resembling chemical kinds. The response to the next objection discusses cases of this form and the options that they generate.

this minimal variability to be sufficient for MR. Rather, multiply realized kinds are said to be "wildly disjunctive" or "heterogeneous."[6] Functional kinds are relevant here because the same functional role can often be occupied by extremely diverse physical mechanisms, both in terms of composition and structure. For example, the same computer program can be implemented in quite diverse physical hardware and by CPUs with different instruction codes. Many will claim that this type of compositional and structural insensitivity is the hallmark of MR. However some properties that exactly resemble at one level of abstraction have variable realization bases, hence meeting our standards for MR, but lack this required insensitivity to composition and structure. For example, a kind might have a very limited range of realizations, or the realizations might compositionally or structurally resemble one another much more closely than do the diverse realizations of a functional kind. Such a kind should not count as multiply realized, this objector argues, but turns out as such according to our analysis.

I have two responses to this objection, but I will also go on to acknowledge that there is some truth to it. First, let us note that judgments of similarity are always relative to some level of abstraction. Functional kinds are distinctive in that they are multiply realizable even at high levels of abstraction. For example, various medium-sized contraptions can occupy the role of a carburetor. This heterogeneity at our everyday, medium-sized level of abstraction makes *carburetor* an obvious example of a multiply realized substantival kind. But note that other kinds that do not have many or diverse medium-sized realizations can still have (and likely do have) diverse, "wildly disjunctive," "heterogeneous" realizations at a much lower level of abstraction. So they still meet this more restrictive standard for MR relative to at least one level of abstraction. We can speak of such kinds as being multiply realized in the microphysical.

But second, the wildness and heterogeneity requirement itself, regardless of the level of abstraction to which it is applied, introduces a vagueness that threatens the connection MR is supposed to bear to ontological reduction. Our analysis provides sharp conditions for MR: any difference in realizations, at some level of abstraction, through sameness at the higher level of abstraction, is sufficient for MR. Even a higher-level kind that is

[6] Kim (1993a: 321).

coextensive with the disjunction of two nomologically possible realization kinds turns out to be multiply realizable on this view. Critics may point out that such sharp conditions cannot possibly be correct, because MR is a vague concept or at least admits of degree. In response, there is still a clear sense in which MR admits of degree on the present analysis. Just as we can speak of sentences being "more disjunctive" in virtue of having more disjuncts, so too we can speak of greater MR in virtue of a kind having more realization kinds. And the dissimilarities of the realization bases, again at some lower level of abstraction, provide another dimension for determining the degree of MR. These dimensions—number of realization kinds and their degree of dissimilarity—are surely relevant to some important senses of 'reduction.' Greater realization heterogeneity suggests greater methodological autonomy for the realized kind.

But why should a higher-level kind that has only two, perhaps closely resembling, realization kinds count as multiply realized? One answer is that MR has traditionally been used as a tool for refuting simple one-to-one kind identities in the philosophy of mind. If *pain* were coextensive with two neuroscientific kinds, then *pain* clearly could not be identified with merely one of these kinds. Compare this with Kim's example of *jade*, which is coextensive with two mineral kinds, *jadeite* and *nephrite*, so not reducible to either in isolation.[7] Perhaps *jade* is a kind, but not a mineral kind—that is, it has different specification dimensions than do the mineral kinds *jadeite* and *nephrite*. In this example, *jadeite* and *nephrite* would not be determinates of *jade*. Perhaps *jade* does not even have any determinates. As we have seen in our discussion of Kim's "disjunctive move," kinds that are coextensive with disjunctions of *Y*-kinds are not necessarily *Y*-kinds themselves. This may be because there are no such disjunctive kinds at all, or there are such kinds but they are not *Y*-kinds because they do not figure in any *Y*-laws (or perhaps any other laws, for that matter, due to projectibility worries) or abide by the *Y*-theory classifications (e.g., specification dimensions).[8] It is epistemically possible (for me, at least) that *jade* is not a mineral kind, but its two mineral varieties—*jadeite* and *nephrite*—are two of its mineral realizations. Likewise, it is epistemically

[7] Kim (1993b: 319–322).

[8] For the former route see Armstrong (1978: ch. 14); for discussion of the latter, see Fodor (1974, 1997) and Kim (1993b).

possible that *pain* is not a neuroscientific kind, in virtue of having multiple (even if only two) neuroscientific realizations.

It should be widely accepted that this liberal conception of MR is sufficient to deny kind-kind identities on the Nagelian, positivistic conception of reduction. However, it might be argued that more sophisticated versions of reduction have been developed in the last few decades that are not refuted by this liberal conception of MR.[9] In response to these more sophisticated versions of reduction, some may argue that a more sophisticated concept of MR is also required. But I do not think that this is so. Even minimal disjunctivity, as in the case of *jade*, forces theoretical decisions concerning whether determination or realization relations hold.

Determination or Specification. The disjuncts might be multiple determinates or specifications of the broader kind. For example, *jade, jadeite*, and *nephrite* might each be accepted as legitimate substantival kinds, with *jadeite* and *nephrite* two of the specifications of *jade*. In order for this to be the case, *jade, jadeite*, and *nephrite* would have to fall under the same specification dimensions and simply differ in the values they span along those dimensions. Or take our familiar example of the adjectival kinds *red, scarlet*, and *crimson*, which fall under the same determination dimensions and differ merely in the range of values they span along those dimensions. This is an example of determination. In these cases the "disjunctive kind" is retained and the "disjunct kinds" are its specifications or determinates. The discovery of these multiple specifications might suggest revising our understanding of the specifiable or determinable kind. We might even uncover the original kind's determination dimensions by first discovering its determinates.

Elimination. The purported kind in question might simply be eliminated. For example, we might conclude that whereas we once thought there is this one kind *jade*, there are actually two resembling kinds—*jadeite* and *nephrite*. However, neither *jadeite* nor *nephrite* is *jade*. Further, in contrast with a retention with determination or specification, *jadeite* and *nephrite* do not share specification dimensions with the supposed kind *jade* (or with each other, perhaps).

[9] For example, see Richardson (1979), Hooker (1981), Churchland (1986), and Bickle (1998, 2003).

In this case, we should recommend, for our scientific purposes at least, replacing our jade talk and jade generalizations with more fine-grained *jadeite* and *nephrite* talk and generalizations. Most likely, such an elimination will occur when there are no generalizations about *jade* "as such."

Revisionary Ontological Kind Reduction. Alternatively, the concept of a particular kind might be revised and that kind then reduced to one of its realization kinds. For example, suppose that instances of both *jadeite* and *nephrite* meet many of the criteria for *jade*, but neither does so perfectly. However, instances of *jadeite* better meet these criteria. Or suppose that jadeite is simply more prevalent than nephrite in the population space. Our concept of jade might then be revised, and we might conclude that *jadeite* is *jade*. This is a revisionary ontological kind reduction. (Such revisionary reductions can also occur when there is only one realization kind.) The vagueness in our concepts, the acceptance of near-fits, and the distribution of samples allow for vagueness about even ontological reduction. But it would be a mistake to think that kinds themselves can be vague. Rather, the vagueness arises in whether two terms or concepts pick out the same kind. It is not a vague matter whether MR—a relation between kinds—ever occurs.

Ontological Kind Autonomy. Finally, the kind in question might be retained, though it neither reduces to any of its realization kinds nor is it determined or specified by them. In this case, the kind in question would operate at a different, but still legitimate, level of abstraction with its own determination or specification dimensions and non-determinable necessities. The difference in abstraction shows that this is a case of realization and not determination. This status would be vindicated by the existence of laws, causal explanations, real similarities, etc. that make essential reference to this kind. This is ontological kind autonomy, and only here do we have MR. For example, perhaps *jade* is not a mineral kind, so it does not reduce to *jadeite* or *nephrite* nor does it share their specification dimensions. But it still is a kind invoked in scientific laws, causal explanations, real similarity judgments, etc. This scenario differs from determination or specification with retention in that, in the present case, *jade*, *jadeite*, and *nephrite* do not all fall under a common determinable or specifiable kind.

In addition to introducing the concept of relative MR, there are other interesting and fruitful ways to classify cases of MR that should be acknowledged when considering the variability of realizations. Prominent among these is a division between *compositional MR* and *structural MR*. A kind is compositionally multiply realizable, I will stipulate, if it has possible realizations that are structurally isomorphic (at some level of abstraction) but differ in material composition. To borrow from Shapiro's example, consider corkscrews that are structurally isomorphic though one is made of steel and the other is made of aluminum. Though Shapiro denies that this is a case of MR, this example meets the present standards for compositional MR. Another familiar example of this type is the piecemeal replacement of a person's neurons with some kind of artificial material, with the hope that this will preserve the higher-level mental properties. If sameness of mentality would be preserved through such replacement, then this too would be a case of compositional MR.

A kind is structurally multiply realizable if it has possible realizations that differ structurally—loosely, in the arrangement of their parts—whether or not there is a compositional difference. Suppose that two animal species can both form the same visual images of their environments, but the realization bases for their visual systems exhibit different structural patterns and employ different engineering strategies to deal with optical problems.[10] This would be a case of structural MR.

It might be that one of these types of MR is more important when it comes to assessing various reductive and non-reductive projects. Shapiro, for example, has discounted the relevance of compositional MR in this regard.[11] Again, for Shapiro mere compositional differences are not different *ways* of performing a function, so they cannot by themselves (i.e., independent of the functional differences they may give rise to) ground MR claims. I stand by the present account of MR according to which mere compositional differences can ground the MR of a functional (or non-functional) kind. Nevertheless, Shapiro very well might be right that the more interesting—diverse, higher-level, etc.—type of MR for a functional kind is structural MR. But, significantly, this is probably because we already *assume* compositional MR for functional kinds—it does not matter what material a functional kind is realized in so long as it gets the

[10] Shapiro (2004: 94–103) discusses this example in detail.

[11] Shapiro (2004: 57–65).

job done. We should remember that Shapiro focuses on functional kinds, but non-functional kinds can be multiply realizable as well. And for kinds that, in contrast with functional kinds, have essences that include compositional features, compositional differences in their realizations (which nevertheless are irrelevant to individuating the higher-level compositional kind) might prove more interesting. For example, if *jade* is a kind that is realized by *jadeite* and *nephrite*, the compositional differences of *jadeite* and *nephrite* might prove more interesting than their structural differences (or, at least, more interesting than compositional differences are for functional kinds). It should not be surprising if a compositional kind (i.e., a kind individuated by its material constituents) has realizations that differ in their structural arrangement. It would be more surprising if realizations with lower-level compositional differences could give rise to the same, maximally specific, higher-level compositional kind.[12]

We can also classify realizations and their features according to the contribution they make to their realized kinds. Inevitably, some realization features will make a more significant contribution to the nature of the realized kind—e.g., they will go further towards establishing its determination dimensions values and non-determinable necessities—than do other features of the realization base. Naturally, we should expect practicing scientists who are studying a higher-level kind to be more interested in those realization differences that make a greater difference when it comes to shaping the nature of the higher-level kind. I believe that Shapiro is close to making this classification of realizations when he distinguishes between those properties of the realization base that are relevant to the purpose of a functional kind and those that are irrelevant. Realization differences that contribute to the capacities definitive of a functional kind are more interesting than those that do not so contribute (or contribute to a lesser degree). But, again, such realization differences should not make a higher-level difference. That would be a case of determination or specification, not realization.

Objection 4. MR is not possible for any kind.

[12] I recognize that there are skeptical considerations against the composition/structure dichotomy that I have presented. I have in mind the type of skepticism similar to that expressed in Lycan (1981, 1990). Lycan argues that the function/structure and software/hardware distinctions are level-relative. (Lycan is obviously using 'structure' in a different sense than we are using it.) I accept this point. Whether we have a case of compositional or structural MR is similarly relative to the level of realization at issue.

On the present view, MR is rampant. At the other extreme, some—such as David Armstrong (1997), John Heil (1999), and Heil and C. B. Martin (1999)—deny the very possibility of MR (in the present sense) by judging a realization difference to be, *ipso facto*, a higher-level kind difference as well. As Heil puts it:

> True enough, the predicate 'is in pain' holds of particular objects at particular times, and it holds of those objects in virtue of their properties. The property in virtue of which an object satisfies 'is in pain', however, is not the generic property of being in pain. There is no such property. . . The idea, rather, is that 'is in pain' applies to creatures that are similar in certain salient respects: similar enough to merit application of the pain predicate. These similarities stem from the creatures' possession of certain properties. The properties need not be the same in every case: creatures are neither identical nor exactly similar in those respects in virtue of which they satisfy the predicate 'is in pain'.[13]

To claim MR for a kind is to claim that the same super-determinate or maximally specific kind can be realized in diverse lower-level mechanisms. But if a difference in lower-level mechanism is, *ipso facto*, a higher-level difference, then there cannot be MR. Claims of MR would then have to be paraphrased away. On this view, to say that psychological kinds are multiply realizable is really to say that there are more psychological kinds than one might pre-theoretically think. Specifically, there is a distinct psychological kind corresponding to each realization kind—e.g., not just *pain*, but *human pain, alligator pain*, and *octopus pain*.[14] Or, perhaps such higher-level kinds should be eliminated or denigrated to "second-class" status. Though they do not explicitly put the point this way these objectors seem to see a contradiction in the very idea of MR, where this is understood in our strong sense of sameness of super-determinate or maximally specific kind through difference in realization kinds. The fundamental idea behind MR is sameness at a higher-level through lower-level difference. But if higher-level properties arise from and are composed of these lower-level properties, these objectors argue, then there is a higher-level difference whenever there is any lower-level difference.

MR would be impossible were higher-level properties composed of the lower-level properties that realize them in the same way that, say, an

[13] Heil (1999: 201).

[14] Here they join Kim, as discussed in §4.4.3. Churchland (1986) and Thomas Polger (2002) also suggest such kind-splitting, as does David Lewis (discussed in §5.4.1).

animal body is composed of its cells *and* these lower-level differences were always relevant to specifying the higher-level kind. But I simply deny both claims. MR is possible precisely because the compositional rules that hold for *concreta* do not hold for properties. This is the very nature of abstraction. Ontological points aside, it should be obvious by now that higher-level kinds can be maximally specified without paying heed to lower-level realization details. For these reasons, MR is at least possible.

5.3 Memory: A Case Study of Fragmentation

I now want to turn to a real-world example of fragmentation that exhibits the contrast between determination and realization, and which reveals the theoretical moves that we found when considering Kim's hypothetical case involving *jade. Memory* serves as a good example, because it is a scientific kind that has fragmented into more fine-grained kinds. But it is not always clear whether these more fine-grained kinds are determinates of a common determinable (*memory*), multiple realizations of some *memory* kind, or distinct kinds with their own property spaces. The fragmentation of *memory*, then, offers a good illustration of the complex interplay between discovering determinates of a kind, multiple realizations of a kind, and new determinables. So it is like Kim's *jade* example, but only more complex and detailed. It is also helpful to look at some generalizations about memory to test Kim's skeptical challenge regarding the possibility of a science of multiply realized kinds.

It is now widely recognized that even when we limit ourselves only to human memory, a great variety of kinds fall under the memory category. Ever since Atkinson and Shiffrin (1968) introduced the modal memory model, it has been common to divide human memory into three broad kinds with their own memory systems: *sensory memory, short-term memory*, and *long-term memory*. Let's begin by briefly characterizing each of these kinds.[15]

Sensory memory is the short-lived retention of information from a sensory impression. The duration of sensory memories varies according to the sense modality, but they never last more than a few seconds. In the case of vision, sensory memories disappear after only 200–300 milliseconds.

[15] Bjork and Bjork (1996) serves as a good introduction to these topics and was used as a reference for much of the discussion to follow.

The most commonly studied sense modalities for memory research are vision and audition—iconic memory and echoic memory, respectively—but there are sensory memories corresponding to the other sense modalities as well. So we can generate a taxonomy, in Figure 5.1, depicting what I take to be determination relations.

Another important feature of sensory memories is that they are not translated into another code, and they retain the phenomenology characteristic of the original sensory experience. As such, sensory memories can be quite vivid, though they fade quickly. Sensory memory is thus short term, but high density. Sensory memories can, of course, also be transferred into our longer-term memory systems. Finally, sensory memory cannot be improved through the use of remembering strategies such as rehearsal.

Short-term memory, or working memory, differs from sensory memory in its duration, capacity, coding, and accessibility. Though different models vary in their estimates, short-term memories typically are retained for no more than thirty seconds to a few minutes. Further, the storage capacity for short-term memories is quite limited. George Miller (1956) famously put the limit at seven items, plus or minus two, though his estimate might be a bit high. This capacity can be increased, however, by "chunking" information into meaningful units—e.g., remembering twenty-one numbers instead of seven by breaking the string of numbers into three-digit chunks. The dominant code for short-term memory is acoustic (phonological).[16] Short-term memories are also easily retrievable, as they are in some sense activated memories. And they can be further retained in short-term memory by rehearsal.

It is not as easy to discover plausible determinates of *short-term memory* as it was for *sensory memory*. Though there are models of short-term memory that aim at uncovering its functional structure. One such model is Baddeley and Hitch's (1974) working-memory system. They posited a three-part working-memory system consisting of the articulatory loop, the visuospatial sketch pad, and a central executive. The articulatory loop is supposed to be the subsystem that translates memories into phonological code and generates subvocal speech (memory rehearsal). It is further divided into two parts—a phonological store and articulatory control process—

[16] Baddeley (1966); Bjork and Bjork (1996: 106–107).

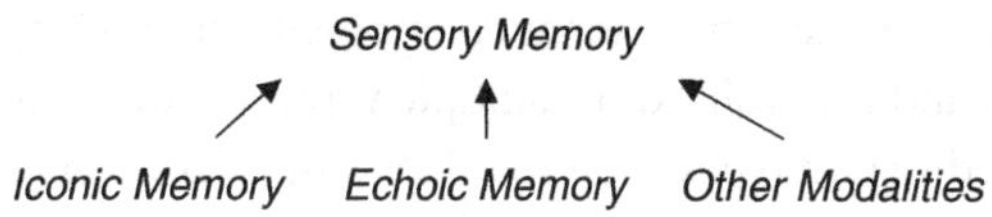

Figure 5.1 *Memory* determinates

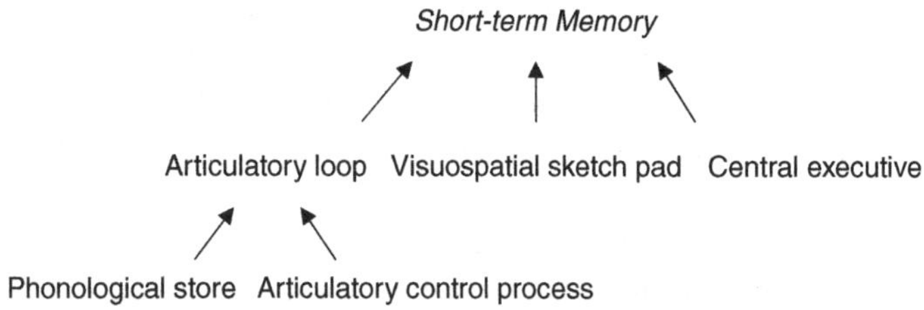

Figure 5.2 Functional analysis of *short-term memory*

corresponding to these two functions.[17] The visuospatial sketch pad is the subsystem that, as the name suggests, handles visual and spatial information. The central executive is then in the administrative position of assigning work to the other two subsystems. This structure is depicted in Figure 5.2.

The structure depicted in Figure 5.2 is different in kind from that depicted in Figure 5.1. Figure 5.2 does *not* depict determination relations. Instead, this figure depicts the different functional components of *short-term memory*. And different functional components of a system are not to be mistaken for determinates of a kind. For example, crankshafts are not determinate kinds of engines; rather, they are functional components of engines. Figure 5.2 offers a functional analysis of *short-term memory* and not a determinable–determinate analysis, even if visual and phonological memories are different kinds (determinates) of *short-term memory*. This figure is included to make this contrast clear. The relevant difference is that Figure 5.1 depicts a taxonomy of *memory* kinds, whereas Figure 5.2 depicts a proposed breakdown of the functional components that give rise to a *memory* kind. Figures depicting the functional components of a system can be thought of as a step toward the depiction of realization relations. This is *only* a step, however, because it is often an open

[17] Bjork and Bjork (1996: 109).

question whether these functional components themselves are distinct mechanisms and are realized in multiple ways. We now should see that at least three different types of figures depicting relationships among distinct kinds can be distinguished. These are figures depicting determination relations, functional analyses of a system, and realization relations.

Some of the best evidence for a systems division between short-term and long-term memory comes from cases of amnesia. For example, patients with anterograde amnesia, like the famous H. M. (Henry Molaison), are able to maintain information over the short term but are unable to retain any new information, especially declarative memories (see below), over the long term. Long-term memories established before the onset of anterograde amnesia remain intact, however. This suggests that the patients have lost the ability to transfer memories from a short-term to long-term memory system. In cases of retrograde amnesia, in contrast, patients are unable to access memories from before the onset of the amnesia. Both conditions show how the functioning of short-term memory and long-term memory can be radically severed, suggesting that they are served by distinct systems. Amnesiac conditions are often caused by damage to the hippocampus, suggesting its prominent role in creating long-term memories (especially declarative memories).

In properly functioning people various factors can cause memories to be transferred from short-term memory to long-term memory. Such transfers are more likely if the information is rehearsed—especially if rehearsed over spaced intervals and with the use of mnemonics—or if the information is particularly vivid (e.g., "flashbulb memories"). Long-term memory differs from short-term memory in various regards. First, long-term memories obviously have a longer duration—they can last as long as a human life. The capacity for long-term memory is also much greater than that of short-term memory, though the limits of long-term memory are not known. Long-term memories also differ from short-term memories in that they are semantically encoded, as opposed to being acoustically encoded. For this reason, recollection mistakes in long-term memory are more likely to occur with items similar in meaning as opposed to items similar in sound. Retrieval is more complicated for long-term memories, as long-term memory has a greater capacity and its memories are not activated. Retrieval can be aided by various retrieval cues, however. Long-term memories are thought to be globally distributed or realized in the brain.

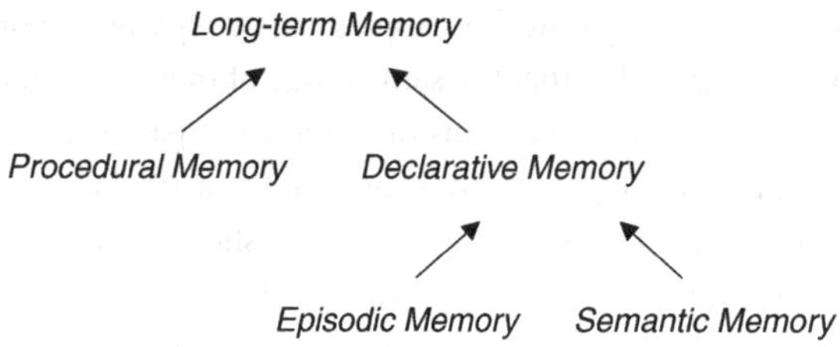

Figure 5.3 *Long-term memory* determinates

Long-term memory itself is often divided into *procedural memory* and *declarative memory*. Procedural memory can be thought of as how-to memory, such as the motor skill memory we might possess with respect to riding a bike. Declarative memory can be thought of as memory-that, such as my memory that Cheyenne is the capital of Wyoming. It is common to divide *declarative memory* into two further kinds: *episodic memory* and *semantic memory*. Episodic memory is essentially autobiographical, whereas semantic memory is not essentially autobiographical. *Procedural memory* also admits of further specification, but for our purposes we can rest with the partial taxonomy provided by Figure 5.3. I understand Figure 5.3 as depicting some of the determinates of *long-term memory*.

The discussion to this point has been intentionally ambiguous in one regard, and that ambiguity should now be clarified. Sensory, short term, and long term can be thought of as different memory systems or as different kinds of memories. A memory system is a system that receives, encodes, stores, and retrieves information. The memory is that which is received, encoded, stored, and retrieved. Obviously, there is a conceptual difference here between memory systems and memories themselves. The ambiguity in the above discussion is that those who distinguish between sensory memory, short-term memory, and long-term memory might simply be positing different memory systems, or they might be distinguishing different kinds (determinates) of memories. Cognitive scientists who endorse the modal memory model view these as three different memory systems. But we could also consider how memories can differ from one another, not how memory systems can differ from one another. So we could inquire after the varieties of memories, rather than the varieties of memory systems.

It is worth considering how these topics can be separated. For example, it certainly seems possible that the same range of memories that we actually see in human beings—in terms of duration, capacity, encoding, and retrieval—could also be possessed by animals or artificial devices with a unitary memory system. The fact that we have short-term and long-term memories does not, by itself, necessitate distinct memory systems for these different kinds (determinates) of *memory*. It just so happens, as an empirical fact, that we likely have different memory systems corresponding to these different portions of the property space of *memory*. But, again, it is at least epistemically possible for there to be an animal or artificial device with a single memory system corresponding to this same property space. *Memory* kinds and memory systems can also come apart in the other direction. There could be distinct memory systems that produce the same kinds (determinates) of *memory*. Such an arrangement would be a bit redundant, however, and our biological reality seems to be more specialized and frugal than that.

We have seen that *sensory memory, short-term memory*, and *long-term memory* are all determinables that admit of specification. And these are all called "memories." But, it would be a mistake to infer, simply on the basis of this linguistic similarity, that *sensory memory, short-term memory*, and *long-term memory* are themselves determinates of a common determinable—*memory (simpliciter)* or *human memory*. Prior to the scientific study of a classificatory system, it is often epistemically possible that the presumed determinates of a kind actually belong to distinct property spaces.[18] That is, for all we know the presumed determinates might possess distinct determination dimensions, though at a superficial level they resemble one another or due to some purely historical accident they have been erroneously grouped together under the same name or category. This might have been, and still might be, the case with *memory*. To justify the claim that a taxonomy represents determinates of a common determinable one should be able to provide their common determination dimensions or otherwise make a case that they do share determination dimensions even if they cannot yet be provided. So, is there a common property space to which these three kinds of memories belong?

[18] But in some cases this is never an epistemic possibility. For example, it was probably never epistemically possible that *scarlet* and *crimson* belong to distinct property spaces.

Recall that we can begin with a rather intuitive test: what are the ways in which memories can differ or resemble *as memories*? What are the differences and similarities that matter to the individuation of memories? Possible dimensions include: duration, coding, and accessibility. Of course we should defer to the experts on memory research, but these are plausible ways in which memories can differ from one another. They also are dimensions along which memories generated by the three different memory systems systematically differ. These are at least plausible candidates for determination dimensions of *memory*. Other plausible candidates likely should supplement this list, though they might not be salient in discussions of memory systems because the memory systems are not divided according to these dimensions. One plausible candidate here is content. Memories can differ in their content, though multiple memory systems might be capable of handling (encoding, storing, and retrieving) memories with the same content.

As with the discussion of *color* earlier, my goal is not to conclusively argue for some set of determination dimensions for a kind. Rather, I intend to use *memory* as a real scientific example in which it certainly seems that determination dimensions guide at least some prominent taxonomies. I would also suggest that it is often advantageous for practicing researchers across the scientific disciplines to have an explicit understanding of the concept of a determination dimension and, in developing or modifying scientific theories, to make the determination dimensions that guide their taxonomies explicit. Not only will this make for a clearer and more orderly presentation of the theory, but it will also suggest possibilities—such as unexplored locations in property spaces or alternative determination dimensions—that might not otherwise be easily recognized.

Recall from §2.2 that determination dimensions often play a prominent role, for example as a variable, in generalizations or laws concerning some network of kinds. With this idea in mind let us consider some of the well-confirmed generalizations concerning human memory, here understood as a memory system:

I. Short-term memory lasts for only about thirty seconds.
II. Short-term memory can hold only about five items at a time.
III. Chunking increases the capacity of short-term memory.
IV. Retrieval from long-term memory is enhanced by spacing rehearsal intervals over time (spacing effect).

V. The earlier and later items on a list are more likely to be remembered than are the middle items (primacy and recency effect, respectively).

VI. Retrieval from long-term memory is enhanced when memory cues match the original encoding (encoding specificity principle). For example, words encoded as pairs are more likely to be remembered when cued by their original pairing.

We see candidate determination dimensions for memory systems playing a prominent role here. Duration is prominent in I; II concerns capacity; III concerns encoding and capacity; and IV–VI concern accessibility and encoding.

An interesting question is whether these generalizations and others like them hold for *memory (simpliciter)* or whether they are specific to *human memory* or some slightly broader biological classification (e.g., *mammalian memory*). If the latter is the case, presumably it would be because the applicability of the generalizations to a population depends on how the memories are realized within that biological population or because the populations simply differ in the kinds of memories they can possess. But if the former is the case, then we have real-world examples of scientific generalizations that hold for multiply realized kinds. In §4.4.3 we were presented with Kim's Challenge to the possibility of such generalizations. But does Kim's skepticism hold up with respect to these six generalizations?

In order to answer this question we need to consider memory systems that differ from human memory systems in their realizations. Nevertheless, these should be memory systems capable of encoding, storing, and retrieving the same kinds of memories as are processed by human memory systems. This is no easy task. One worry is that few non-mammalian animals, let alone actually existing artificial systems like computers, can instantiate the same super-determinate kinds of memories that human beings can instantiate. This is because these non-mammalian animals and computers do not possess the same sensory inputs, interconnections with other cognitive functions, etc. that might be essential to such memories. Let's focus on the computer possibility. To overcome the deficits of actual computers, let's simply imagine a real technological advance with respect to the ability of artificial computers to mimic human sensory inputs and human cognition. Assuming that such an alternative memory system can be manufactured or simply imagined, we can investigate whether the above memory generalizations hold for it.

For radically different realizations of, say, *long-term memory*, it seems unlikely that these generalizations will carry over. Artificial memory systems, like those found in computers, present possibilities for radically different realizations of *memory*. As with human memory, computer memory is divided into distinct subsystems. Roughly, the primary storage of a computer (e.g., cache and RAM) is analogous to short-term memory or working memory, and the secondary storage of a computer (e.g., hard disk drive) is analogous to long-term memory. The analogies hold along multiple dimensions. Primary storage typically has much less capacity than does secondary storage, and it is also more easily lost. This is also the case for our short-term and long-term memory. Primary storage is directly connected to the CPU, and human short-term memories are in an activated state. In contrast, memories in secondary storage must be retrieved by input/output channels in order to reach the CPU, and memories in human long-term memory must be retrieved in order to be activated. Though there are these analogies, we should be leery of accepting the claim that existent computers and human beings possess memories in the very same sense (i.e., that they can instantiate super-determinates, or even close approximations, of the same *memory* kind).

As a feat of imagination, let's conceive of an artificial computer that can at least approximate memories that actual human beings have stored in long-term memory. Let us focus on semantic memories, such as the fact that Cheyenne is the capital of Wyoming. Imagine that both our computer and human being acquired this information by reading an alphabetical list of the fifty states matched with their capitals. The computer then stored this information on the magnetic material covering some platter in its hard disk, and the human being stored the information in her own way. Let's now consider our retrieval generalizations, IV–VI, provided above. Should we expect these generalizations to hold for the computer? I think this is very doubtful.

Generalization IV states that retrieval is aided by spacing rehearsals over a period of time. So our human being will have a better chance of remembering the capital of Wyoming if she were to study her list for three different ten-minute sessions, rather than if she were to study the list for thirty minutes in one session. Is this true for the computer also? No. The information is stored on the hard disk drive the same whether it is inputted in rapid succession or over an elongated period. Further, the spacing of input (rehearsal) does not alter the effectiveness of the busses that connect the hard disk drive to its CPU (retrieval).

According to generalization V our human subject should have an easier time remembering the capitals of Alabama (primacy) and Wyoming (recency) than she will remembering the capital of Montana. But the same generalization does not hold for computers accessing their secondary storage devices. Nor is a computer's ability to retrieve information typically aided by memory cues that match the encoding environment. Though, it might be easier for a computer to retrieve information in chunks that were stored at the same time and location.

It should be clear that these generalizations about *memory* are not generalizations about *memory (simpliciter)*, but about memory as realized in human memory systems. This point holds for generalizations I–III as well. This is consistent with Kim's skepticism, though there might be interesting generalizations about *memory* "as such" that I have simply not presented. Regardless, it seems that at least many of the *memory* generalizations are restricted to particular realizations. This suggests a reductionistic methodology—to fruitfully practice the science of *human memory* we need to do the science of the realizations of *memory* in humans.

How should we handle our memory taxonomies when we discover that our generalizations about *human memory* do not carry over to other apparent memory systems—e.g., artificial computers or different biological species? There are three main options. First, we could admit that these are still memory systems of a common determinable kind. But it is difficult to reconcile this with the fact that the same generalizations do not hold for these other systems that, by our suppositions, seem to handle the same kinds of memories as do human memory systems. It seems that we would have to claim that the memory systems of humans, other biological species, and artificial computers differ along hitherto unknown determination dimensions that they nevertheless share. Second, we could use the fact that different generalizations hold for these different memory systems, though the distinct systems share the same determination dimension values *qua* memory systems, to show that these are not really *memory* generalizations after all. Instead, they are realization-specific generalizations. One then worries, however, whether there are any generalizations about *memory (simpliciter)*. If not, the status of this supposed scientific kind is put in serious doubt, and the next option becomes more likely. Third, we could conclude that the memory systems of humans, other biological species, and artificial computers belong to distinct property spaces. There is no single determinable kind that these systems all manifest, even though the same term, 'memory,' applies to each. This is an eliminativist position with respect to the supposed scientific kind *memory (simpliciter)*.

5.4 Explanatory Benefits of the Present Analysis

We now have an account of MR with concrete and detailed applications. But we cannot be confident that we have a correct, or better yet profitable, analysis of MR until we use that analysis to show how the concept of MR can do much of the theoretical work that it is supposed to perform and that is worth performing. This work was detailed in §4.3, where arguments for three senses of 'autonomy,' from premises concerning MR, were presented. It is now time to show that the present analysis of MR fits this theoretical role. I will do this by arguing that the present understanding of MR fits the needs of the first premises of each of these §4.3 arguments. I will not be concerned with defending the second premises of these arguments—the premises concerning the reality of MR for some domain. For present purposes I am interested in clarifying the concept of MR and its implications rather than establishing that some particular kinds actually are multiply realizable. This is because the first premises of these arguments fall within the domains of metaphysics or philosophy of science, but the second premises are within the domain of various conceptual schemes and sciences. As such, the experts within those domains should determine the truth or falsity of the second premises.

5.4.1 *Weak Ontological Autonomy*

Our MR Argument for Ontological Autonomy purports to establish that multiply realizable kinds do not reduce to their realization kinds. The idea is that multiply realizable kinds cannot be identified with their realizations individually or via logical operations like disjunction. The conclusion of this argument is very similar to the conclusions reached by Putnam and Fodor in their early arguments for functionalism. For example, consider the following passage from Fodor:

> To put it briefly, type identical events do not, of course, have *all* their properties in common, and type distinct events must nevertheless be identical in *some* of their properties. The empirical confirmation of token physicalism does not depend on showing that the neurological counterparts of type identical psychological events

are themselves type identical. What needs to be shown is just that they are identical in respect of those properties which determine what kind of *psychological* event a given event is.[19]

Here we see Fodor claiming that neurological differences are not necessarily psychological differences. Neurological kinds and psychological kinds have distinct individuation conditions (e.g., determination dimensions), so psychological kinds cannot be identified with neurological kinds. The last sentence of this quotation foreshadows our distinction between determination and realization. Different realizations of a psychological kind are not identical *simpliciter* (naturally), but "they are identical in respect of those properties which determine what kind of *psychological* event a given event is." The "properties" that determine the kind of psychological event at hand are what we have called the determination or specification dimensions for that psychological kind. Fodor's italics correctly suggest that when classifying psychological kinds all other properties should be abstracted away.

Fodor's discussion of different neurological kinds that realize the same psychological kind contrasts with our discussion of *red* and its various shades. *Crimson* and *scarlet* are not multiple realizations of *red* because, to paraphrase Fodor, they are not identical in respect of those features which determine their *color* kind. So, the critical question for determining the MR of an *X*-type is captured by our analysis of MR: do the realization differences mark a difference in *X*-ness?

It should be obvious that the present account of MR supports Putnam's insight that MR is the downfall of kind identity theories. If *pain* is multiply realized with respect to neuroscientific kinds, then according to our analysis there is more than one neuroscientific kind that can serve as the realization base for the same super-determinate *pain* kind. As no particular neuroscientific realization kind is necessary for that super-determinate *pain* kind, it is not possible to reduce, or at least identify, that *pain* kind to any particular neuroscientific kind.

Whether the present account of MR allows for token reductions of multiply realized kinds is a more complicated matter. A token reduction obtains whenever instances of some kind *A* are identical to instances of some kind *B*, regardless of whether kind *A* is identical to kind *B*. The

[19] Fodor (1974: 126).

strong version of ontological autonomy, in addition to denying kind-level reductions, also denies such token reductions of a multiply realized kind to its realizations. To settle the question of token reduction we first need to clarify the ontological status of the tokens in question. Are they events, states of affairs, tropes, or of some other ontological kind? Events are customarily taken to be the relevant tokens (e.g., in the Fodor quotation above or in Davidson's theory of Anomalous Monism), and MR is customarily taken to be compatible with a token identity holding for an *A*-event and a *B*-event, where *B* is one of the realization kinds of *A*. For example, it is common for type non-reductionists to hold that though *pain* is not type-identical to neuroscientific kind *N*, a *pain* event can still be identical to an *N*-event.

Whether such an event identity holds depends on how events are understood. The Davidsonian tradition, which sees events as unstructured, concrete particulars, is conducive to this result.[20] It should not be a surprise, then, that Donald Davidson's Anomalous Monism is the paradigm of a type non-reductionist theory coupled with a token reductionism. I am willing to accept this conclusion, so long as events are construed, in Davidson's sense, as concrete particulars. The possibility of token reductionism is put in serious doubt, however, by alternative conceptions of events according to which they possess an essential/accidental property structure. It is then plausible to hold that the psychological and neuroscientific events in our example differ in their essential properties and, therefore, are not even token-identical.

Back in Chapter 2, I argued that realized properties are not identical to their realization properties. So, MR is not consistent with property identities. The present theory then supports the conclusion of strong ontological autonomy, assuming that the relevant tokens are properties (again, understood as particular instances of kinds). If psychological kinds are in fact multiply realizable, the metaphysics of causation then becomes critical for resolving debates over mental causation. For example, if properties are the causal *relata*, then psychological properties do not become causally efficacious simply because the properties which are their realizations are causally efficacious (as they are distinct). In fact, I accept properties as the causal *relata*. This combination of views—property distinctness

[20] Davidson (1980).

and properties as the causal *relata*—rules out reductive solutions (such as David Robb's) to the problem of mental causation, on the assumption that psychological kinds are multiply realizable.

The connections between properties and causation are extensive. If properties are the causal *relata*, then discovering novel causal relations can be a way of discovering new properties. This connection should be unsurprising. For both causation and properties, similarity or resemblance relations are central. Causal relations are often systematic, such that one category of resembling things is reliably connected to another category of resembling things. Nomological analyses see this feature as essential to causation. Whether or not they are correct, it is at least a commonplace feature of causation in our world. Resemblance is always along some dimension—e.g., determination dimension. Such dimensions, in conjunction with non-determinable necessities, are constitutive of kinds. The discovery of a causal relation could be novel because it involves the discovery of a new kind of resemblance—a new way of grouping things and seeing patterns.[21] And the discovery of a new kind of resemblance goes hand in hand with the discovery of new properties.

Causes, as properties, are individuated by the conditions offered back in §2.7. Consider the following example adapted from Yablo (1992). Sophie is a pigeon that has been trained to peck at all and only red things. A red, wooden triangle is placed before her. Sophie pecks. What caused the pecking? No matter what account of the causal relation one favors, it is clear that the pecking was caused in virtue of a certain abstraction—the redness of this triangle. The property causation proposal holds that this *red* property caused this pecking. All the causally irrelevant properties—e.g., the triangular shape and wooden material constitution—are abstracted away and are literally not a part or constituent of the cause of the pecking. This *red* property has its nature exhausted by its determination dimension values and non-determinable necessities.

While the conclusion of our MR Argument for Ontological Autonomy, the kind-level autonomy expressed by OC, is widely accepted by philosophers of mind, there have been some prominent dissenters. There has been dissension over whether psychological kinds are multiply realizable,

[21] Of course, many novel discoveries of causal relations simply involve known resemblances that were previously not thought to correlate. The sense of novelty discussed in the main text is deeper than this.

as well as to whether kind-level autonomy follows from this (i.e., OP1 is true). Though I have often written on the assumption that at least some psychological kinds are multiply realizable, I do not aim to take a position on that issue here. Rather, I am attempting to settle the more fundamental issue of what is required, in general, for the MR of a kind and what consequences follow from MR. I assume that non-philosophers will then provide the material to settle whether psychological kinds meet this requirement. That being said, I now wish to consider some moves that have been made to deny kind-level ontological autonomy for psychological kinds. The primary reason for doing so is that these responses represent more general strategies that one could take to resist conclusions regarding kind-level ontological autonomy, in any domain, from considerations of supposed MR.

Some have denied the conclusion by arguing that even if psychological kinds are multiply realizable, they are nevertheless related to their realization kinds in a way that deprives them of kind-level ontological autonomy. I know of at least two ways in which this strategy has been pursued. First, some have claimed that multiply realizable kinds are identical to the disjunction of their realization kinds. Hence, they are not ontologically autonomous with respect to their realization kinds. We saw this maneuver when discussing Kim (§4.4.3), and we now know how to utilize the distinction between determination and realization to refute this suggestion. Without question, some kinds are identical to disjunctions of other kinds. *Red*, for example, is identical to the disjunction of all its determinate shades. This is because they all share the same individuation conditions—determination dimensions and non-determinable necessities—and the determinate shades collectively exhaust the range of determination dimension values for *red*. Accordingly, *red* is not ontologically autonomous with respect to its various shades. But let us now assume that *pain* is multiply realizable with respect to neuroscientific kinds. Then *pain* cannot be identified with the disjunction of its neuroscientific realization kinds, because *pain* and these neuroscientific kinds do not share the same individuation conditions (e.g., determination dimensions). The range of determination dimension values for *pain* cannot be constructed from the determination dimensions of its realization kinds since phenomenological intensity, say, is simply not present as a determination dimension for any of these neuroscientific kinds (we are supposing). Further, no instance of *pain* is identical to any instance of these neuroscientific realization kinds. Since *pain* and its neuroscientific realization kinds differ in their determination dimensions, their instances will therefore always differ as well. Recall from Chapter 2 that the nature of a property is exhausted by its determination dimension values and non-determinable necessities.

A second way to deny kind-level ontological autonomy is to argue that psychological terms are non-rigid designators that refer to distinct realization kinds (or their instances) in different contexts. The most prominent advocate of this proposal has been David Lewis.[22] On this proposal 'pain' does not refer to one kind, but to distinct kinds in different contexts. When applied to humans it refers to *human pain*, when applied to reptiles it refers to *reptile pain*, etc. Even greater specificity might be required, as perhaps 'pain' refers to distinct kinds for different individuals of the same species or even to different kinds for the same individual at different times. This proposal, most clearly in its Lewisian manifestation, takes psychological kinds to be second order—the nature of such kinds wholly consists in being of some first-order kind or other, where the first-order kinds are picked out by a functional description. Each of the first-order kinds, in common philosophical parlance, is a realization of the psychological kind.

This proposal bears some similarities to Kim's disjunctive move. But instead of identifying a general psychological kind (e.g., *pain*) with the disjunction of its realization kinds, on this proposal a psychological kind is of a higher order than its realization kinds and wholly consists in being one among these disjuncts (realization kinds). Significantly, the second-order kind is *not* identified with the disjunction of its realizations. Instead of identifying *pain* with the disjunction of its realization kinds, Lewis held that *pain* is identical to different kinds in different contexts.[23] 'Pain,' on Lewis's interpretation, comes across as a slightly odd kind term in that it does not pick out a unique kind. As Lewis puts it:

> If *M* means 'the occupant of the *M*-role' and there is variation in what occupies the *M*-role, then our psychophysical identities need to be restricted: not plain *M=P*, but *M*-in-*K=P*, where *K* is a kind within which *P* occupies the *M*-role. Human pain might be one thing, Martian pain might be something else.[24]

[22] See his "An Argument for the Identity Theory," reprinted in Lewis (1986), as well as "Psychophysical and Theoretical Identifications" and "Reduction of Mind," reprinted in Lewis (1999).

[23] Lewis speaks of 'states' instead of 'kinds,' but this difference is irrelevant. He too was referring to kinds (property types).

[24] Lewis (1999: 305–306).

Lewis differs from Kim in that he is unwilling to identify *pain* with the disjunction of the realizations of, as he puts it, the *M*-role (i.e., the functional role associated with some psychological predicate). But in denying this identity he shares an assumption with Kim. Namely, both hold that the pains of all the actual and possible animals are a heterogeneous lot. That is, neither Kim nor Lewis holds that there is a "non-disjunctive," simple commonality shared by all creatures that are in pain.

> But don't we think that there is *one* property of being in the state *M*—one property that is common to all, actual or possible, of whatever kind, who can truly be said to be in state *M*?—There is. It is the property such that, for any possible *X*, *X* has it just in case *X* is in the state that occupies the *M*-role for *X*'s kind at *X*'s world. The gerund 'being in *M*' can be taken, at least on one good disambiguation, as a rigid designator of this property. However, this property is not the occupant of the *M*-role. It cannot occupy that or any other causal role because it is excessively disjunctive, and therefore no events are essentially havings of it... Since the highly disjunctive property of being in *M* does not occupy the *M*-role, I say it cannot be the referent of *M*.[25]

In the end, Kim and Lewis both opt for species-specific (or structure-specific, etc.) kind-level reductions and simply differ over their account of the supposedly general (e.g., species-neutral) psychological kinds (e.g., *pain*).

Using our vocabulary, those who endorse Lewis's proposal implicitly recognize that generic psychological kinds and their realization kinds differ in their determination dimensions. This implicit recognition consists in their judgment that generic psychological kinds are second-order and not to be identified with the disjunction of their realization kinds. But it is difficult to see how these second-order kinds are kinds at all, given their disjunctive nature (e.g., determination dimensions). Lewis further deflates second-order kinds by denying that they are ever causally relevant. Whenever it appears that they might be so, it is really the first-order realization kinds that are causally relevant. Given their supposed causal inefficacy and extreme heterogeneity, one might easily be led to deny that there are such second-order kinds at all. Instead, one could hold that there are merely non-rigid (or second-order) predicates that pick out different kinds in different contexts. The general kind *pain*, for example, would then be eliminated and replaced with more specific *pain* kinds that are

[25] Lewis (1999: 307).

identified with realization kinds specific to a species, individual, or even individual-at-a-time.

My account of kinds and MR offers a response to Lewis that is very similar to the response offered against Kim. Against Kim, I objected that *pain* has determination dimensions that are not shared by the disjunction of its neuroscientific realizations. (Of course, this objection does critically depend on a substantive assumption about the nature of *pain*.) Instead of identifying *pain* with a disjunction of neuroscientific realizations, Lewis identifies *pain* with different individual neuroscientific kinds in different contexts. Still, *pain* cannot be identified with any of these individual neuroscientific kinds, even putting aside concerns about non-rigidity, because *pain* does not share determination dimensions with these neuroscientific kinds. They simply are at different levels of abstraction. And so too for various other psychological kinds.

Lewis also fails to allow for the robust kind of MR that I have asserted is at least possible—namely, exact resemblance of psychological kind through difference of neuroscientific kind. For, I have claimed that it is possible for some super-determinate kind of *pain*, say $pain_{1942}$, to have distinct neuroscientific realizations, say NS_1 and NS_2. Whether this is true of *pain* is actually irrelevant in objecting to Lewis, so long as these relations hold with respect to some psychological kinds. Let us suppose that $pain_{1942}$ is realized in Canadians by NS_1 and in Australians by NS_2. Then, if we adopted Lewis's account, we would have to conclude that $pain_{1942}$-in-Canadians = NS_1 and $pain_{1942}$-in-Australians = NS_2. But something has gone wrong here, as $NS_1 \neq NS_2$. The pains are supposed to exactly resemble, but the neuroscientific kinds do not exactly resemble. So, either their pains differ after all or Lewis's reductive account is mistaken. And I see no reason to think that their pains must differ. Lewis, like Kim, makes the mistake of confusing the heterogeneity of *pain*'s realizations with the heterogeneity of *pain* itself.

I have now defended OP1—which connects MR to kind-level ontological autonomy—against Kim and Lewis's challenges. Others who carry the reductionist banner often seem upset by MR arguments for ontological autonomy, yet appear to deny neither the premises nor the conclusion of our argument. I have in mind, in particular, Robert Richardson (1979), Clifford Hooker (1981), and John Bickle (1998, 2003). They argue for theories of reduction that purport to be compatible with kind distinctness. Bickle, in particular, argues that the canonical examples of reductions offered by scientists often do not contain inter-theoretic kind identities (i.e., the concepts of the reducing and reduced theories need not refer to the same kinds).[26] This is fine so far as it goes. But these points are non-sequiturs when it comes to debates over kind-level ontological autonomy. I only mention this reaction because I often get the sense that such "new-wave" reductionists conceive of themselves as objecting, in some sense, to arguments like our MR Argument for Ontological Autonomy.

Of course one could also deny the conclusion of kind-level ontological autonomy by denying the MR claim itself. It might prove insightful to see how this has been pursued by actual disputants in MR debates over psychological kinds. There are always two logically possible ways to go about denying an MR claim. MR requires sameness through difference, so one could deny the supposed sameness of realized kind or deny the supposed difference of realization kinds. Taking the former option, one could argue that the alleged multiple realizations of the same kind are actually realizations of distinct kinds. On the latter option, one could argue that the alleged multiple realizations are actually of the same realization kind. Both options have been pursued with respect to psychological kinds.

In various ways, Thomas Polger (2002), William Bechtel and Jennifer Mundale (1999), and Shapiro (2004) challenge the common assumption that psychological kinds are multiply realizable. Polger has challenged the common assumption that biologically diverse creatures can be of the very same psychological kind. His emphasis on precise sameness of kind shows an appreciation of the role determination plays in MR debates. MR requires sameness of super-determinate kind through realization differences. In effect, Polger has questioned whether *human pain* and *octopus pain*, to use an example from Putnam that he discusses, have overlapping super-determinates. He speculates that *human pain* and *octopus pain* are distinct psychological kinds, though, to use our vocabulary, he judges that they are also determinates of a common determinable. Responses of this form always depend on showing that the supposed multiple realizations of *K* actually differ with respect to their *K*-ness. So for this response to be effective it must be the case that *human pain* and *octopus pain* differ in the "pain-ness."

[26] See, for example, Bickle (2003: 21–26).

Bechtel and Mundale, as well as Shapiro, focus on the realizations and argue that they often do not differ to the extent required for MR. Bechtel and Mundale claim that the plausibility of the MR of psychological kinds is due to the use of individuation conditions with different grains of specificity when distinguishing psychological and neuroscientific kinds. For example, a quite general psychological kind, such as *pain*, is said to have diverse realizations in humans and octopuses. But Bechtel and Mundale claim that if the same grain is used for specifying both kinds, then this appearance of MR evaporates. They argue that humans and octopuses that are in pain have common realizations at a quite general level of abstraction. They admit that at other levels of abstraction these cross-species realizations are of different kinds, but they argue that at those levels of abstraction their pains are of different psychological kinds as well. To use our terminology, they join Polger in claiming that humans and octopuses likely do not instantiate the same super-determinate psychological kinds. Bechtel and Mundale go on to explain the correspondence between psychological and neuroscientific classifications by arguing that considerations of functional role and cross-species comparisons guide neuroscientific taxonomies. If psychological kinds have functional essences that can be satisfied by members of distinct species, it is then no surprise that there is a correspondence between neuroscientific and psychological classifications at a common level of abstraction. Finally, we have already seen that Shaprio has argued that the MR of a functional kind requires multiple "ways" of fulfilling that function. So, for him not just any realization difference entails MR. I have already addressed this claim.

In the last three paragraphs I have made some points that suggest how debates over the MR of some kind will proceed. Again, for present purposes I am not primarily interested in, nor am I equipped to rigorously defend, the MR of any particular kinds. The more important objective of this section has been to defend the claim that if a kind is multiply realizable then, contrary to the positions of Kim and Lewis, that kind is ontologically autonomous with respect to its realization kinds.

5.4.2 *Weak Explanatory Autonomy*

According to EP1 of our MR Argument for Explanatory Autonomy, explanations utilizing terms for multiply realizable kinds are distinct from and irreducible to explanations utilizing terms for their realization kinds. So, how does the distinction between determination and realization bear on weak explanatory autonomy? If determination is confused with realization, unfortunately the connection between MR and explanatory autonomy is lost. This is because explanations utilizing terms for determinable kinds are not autonomous with respect to explanations utilizing terms for their determinate kinds (and vice versa). For such explanations operate at the same level of abstraction (determination dimensions) and differ only in their specificity at that level of abstraction. An explanation that essentially makes reference to an object's redness and an explanation that essentially makes reference to an object's scarletness are not explanatorily autonomous—they both latch onto the hue, brightness, and saturation of an object as being explanatorily salient. These explanations simply differ in their specificity along those dimensions.

That being said, an explanation citing a determinable kind can nevertheless compete with an explanation citing one of its determinate kinds. This is because specificity is itself a standard for evaluating explanations. In general, a good explanation should be specific enough to capture those features that are both sufficient and necessary to account for the *explanandum*. Too much specificity leads to a failure on the necessity standard, and too little specificity leads to a failure with respect to sufficiency. But explanatory competition between determinable and determinate kind terms is a fight within the family, so to speak. It would be inappropriate to refer to an object's redness in explaining why the girl selected it, if she has a preference for only scarlet objects. Conversely, it would be inappropriate to refer to the scarletness of the object if she is attracted to all red objects. But again, 'red' and 'scarlet' are terms that pick out kinds operating at the same level of abstraction. When it comes to explanations, a failure of degree is not so bad as a failure of kind.

The present account of MR stresses that realization differences do not necessarily result in differences at the level of the multiply realized kind. Those who do not distinguish between determination and realization fail to capture this fact. Putnam, in his well-known passage about the square peg, points out that singular explanations should meet the standards of relevance set by the *explanandum*. Again, we can think of the relevant factors as roughly the conditions that are necessary and sufficient in the circumstances to account for the *explanandum*. Given the types of questions that we are typically interested in, the relevant factors in this sense often will be invariant through a variety of microstructural changes across nomologically possible worlds (e.g., "The same explanation will go in any world (whatever the microstructure) in which those *higher level structural features* are present."). And, on the present account, invariance of a kind through variation in that kind's realization bases is the mark of MR.

We can use our theory of kinds to distinguish three interestingly different types of paired and possibly competing explanations. Suppose we are looking for explanations of Jack's wincing behavior.

1. **Explanations with Terminological Differences:**
   A. Jack is in pain.
   B. Jack est dans la douleur.
2. **Explanations with Determination Value Differences (Specification):**
   A. Jack is in pain.
   B. Jack is in pain-state$_{1204}$.
3. **Explanations with Determination Dimension Differences (Realization):**
   A. Jack is in pain-state$_{1204}$.
   B. Jack is in neuroscientific-state$_{1942}$.

When we have only terminological differences, as in 1A and 1B, the explanations often compete in only a very weak sense, if at all. For example, one explanation might be grammatically inferior (e.g., 'Dude, Jack in pain!') or inappropriate in a conversational setting because it is overly formal or flamboyant (e.g., 'Indeed, Dr Jack Sprat is in pain at this precise moment in time.'). But these are merely rhetorical differences and they offer the very same explanation only in different language. Our explanations 1A and 1B, as they are on par on these grounds and are simply translations of one another, obviously do not compete. Of course someone could be *confused* and think that 1A and 1B are distinct explanations, but this does not mean that they really are distinct explanations. The person who understands each can see that they offer the same explanation, in that the two communities of speakers are picking out the same kind and associating the same determination dimensions with that kind.

This situation can also arise within a language when different terms refer to the same kind. This is found in computer science. At the most basic programming level, the machine language of your personal computer consists of strings of binary digits which make up the data and instructions to be retrieved, decoded, and executed by the central processing unit (CPU). Programming in machine language, or instruction code, would be incredibly tedious, and computer programmers seldom operate at this level.

However, in the early days of computing this was the only method of programming available. Significantly, machine languages are tailored to CPUs such that each family of processors possesses a unique machine language. CPUs can directly execute machine language instructions without requiring any translation. Here we find the line dividing hardware and software. In principle, a CPU could be constructed such that any language, even paradigm examples of high-level languages like C++, becomes its machine language. In such a way, we can move the line dividing hardware and software. We do not manufacture such CPUs because it is simply not practical to devise CPUs that can directly execute those commands (e.g., those of C++). Such a CPU would be overly complex and specialized.

Assembly languages are a bit more manageable. Assembly languages employ codes that are more intuitive for human programmers, such as add, sub, mult, and div for the basic operations of arithmetic (as in the assembly language MIPS). In a pure assembly language the instructions correspond one-to-one to machine language instructions, and in this sense assembly languages are tailored to particular machine languages and CPUs. Assembly language is translated into machine language by programs called assemblers. In effect, these assemblers provide us with the one-to-one bridge laws that are characteristic of an inter theoretic and kind-level reduction. Assembly language instructions are implemented by CPUs via their machine language translations, and there is a one-to-one correspondence between assembly language and machine language. It is natural, then, to think that assembly languages are singularly realizable by their machine languages.

Let us get a concrete example of an assembly language translation, for the Intel IA-32 family of processors, before us:

For example, the instruction code sample

```
55
89 E5
83 EC 08
C7 45 FC 01 00 00 00
83 EC 0C
6A 00
E8 D1 FE FF FF
```

can be written in assembly language as follows:

```
push %ebp
mov %esp, %ebp
```

```
sub $ox8, $esp
movi $ox1, -4(%ebp)
sub $oxc, %esp
push $oxo
call 8048348
```

Instead of having to know what each byte of instruction code represents, the assembly language programmer can use easier-to-remember mnemonic codes, such as push, mov, sub, and call, to represent the instruction codes.[27]

Here we see the one-to-one correspondence to machine language that is characteristic of assembly languages. Given this correspondence, we can think of assembly language instructions as simply notational variants of their corresponding machine instructions. To put this in a broader perspective, we can see this as an instance of two concepts or terms picking out the same adjectival kind, in this case a computational operation.

The explanations we find in 2A and 2B, in contrast, do offer different explanations in that they invoke kind terms that pick out different kinds (*pain* and $pain_{1204}$). Let us suppose that 2A and 2B are both true and that each explains, at least to some extent, Jack's wincing behavior. Despite their distinctness these two explanations are not even weakly explanatorily autonomous with respect to one another, I claim, because the two explanations operate at the same level of abstraction and are recognized as such. Again, we understand "same level of abstraction" as the sharing of determination dimensions and non-determinable necessities. Given this shared level of abstraction, the two explanations pick out the same property of Jack's situation and differ only in the specificity with which they pick it out. Further, the competent speakers in this linguistic community recognize, at least tacitly, that the explanations differ only in their specificity (e.g., they can recognize that $pain_{1204}$ is a type of *pain*, and they would associate the same determination dimensions with each). Again, we should allow that candidate explanations referring to kinds related by determination can nevertheless compete in that one might pick out the relevant pattern better than does the other candidate. For example, if Jack responded by wincing only when in the more specific pain-state$_{1204}$, then 2B would better explain his wincing behavior than does 2A. It very well might be, though, that 2A offers an adequate explanation of Jack's behavior. But even though 2A and 2B offer distinct explanations, 2A and 2B are

[27] Blum (2005: 11).

not explanatorily autonomous. 2A is not explanatorily autonomous with respect to 2B, because 2B is simply a more precise statement of 2A at the same level of abstraction and is recognized as such.

It is only with cases like 3A and 3B that we reach the point at which the explanations are weakly explanatorily autonomous with respect to one another. The explanation offered by 3A is autonomous with respect to the explanation offered by 3B, because the property it picks out—Jack's $pain_{1204}$—possesses determination dimensions that are not shared by the property picked out by the 3B explanation—Jack's neuroscientific-$state_{1942}$. Further, this difference in determination dimensions is recognized, at least tacitly, by experts within that linguistic community. The kind *neuroscientific-state*$_{1942}$ is blind to the determination dimensions of *pain*$_{1204}$, and vice versa. So there is no way to capture the relevant psychological generalization (e.g., Jack winces when he's in pain-$state_{1204}$) using the vocabulary of neuroscience.

Returning to the computer science example, assembly languages differ from truly high-level computer languages, such as C++, in at least two important ways. First, the instructions of high-level languages do not correspond one-to-one to machine instructions. Second, high-level instructions are platform-independent, meaning that they can be run on various CPUs. These two claims, which are related given that different brands of CPUs differ in their machine instructions, are also marks of MR. For example, we have continued to assume that *pain* does not correspond to simply one neuroscientific kind. Likewise, a C++ instruction does not correspond to simply one set of machine instructions. And just as *pain* can be multiply realized in creatures with various biological hardware—such as human beings and octopuses—high-level computer languages can be multiply realized in various hardware platforms.

The instructions of a high-level language are often compiled into machine language. Here is an example that, contrary to the assembly language case, shows the one–many relationship that is typical of high-level and machine language instructions:

> The compiling step converts the text programming language statements into the instruction codes required to carry out the application function. Each of the HLL [high level languages] lines of code are matched up with one or more instruction codes pertaining to the specific processor on which the application will run. For example, the simple HLL code
>
> ```
> int main ()
> {
> ```

```
int i = 1;
exit (0);
}
```

is compiled into the following IA-32 instruction codes:

```
55
89 E5
83 EC 08
C7 45 FC 01 00 00 00
83 EC 0C
6A 00
E8 D1 FE FF FF
```
[28]

This same high-level language code would be compiled into different instruction codes for different families of processors. The comparison to, say, a multiply realized cognitive psychological kind should be obvious.

The instructions of assembly languages are singularly realized and are simply notational variants of machine language instructions. This is like two terms or concepts of a scientific or folk theory converging on a common kind. Here we have ontological and explanatory reduction. (Assembly language might be methodologically autonomous, however, given the more user-friendly nature of its mnemonics.) The instructions of high-level languages, in contrast, are multiply realized in machine language in the sense that the very same high-level instruction(s) can be implemented in virtue of distinct sets of machine language instructions. To this extent, high-level languages stand to machine languages as many think that cognitive psychological kinds stand to neuroscientific kinds. Just as the same high-level language can be compiled into distinct machine languages for distinct families of CPUs, it seems to many that the same cognitive psychological kind can be realized in distinct neuroscientific kinds for distinct biological species. High-level computer languages provide us with actual examples of scientific kinds that are ontologically, explanatorily, and methodologically autonomous with respect to their machine language compilations. An industry of computer programmers attests to the reality of such autonomy.[29]

[28] Blum (2005: 8).

[29] Though there are the above similarities, a couple of important disanalogies between the high-level language and cognitive psychology examples should be made explicit. First, in the computer language cases the kinds are actually instructions. The typical kinds posited by the natural sciences, however, are states or capacities of a system. A second difference, which is a corollary of the first, is that the kinds of a high-level language have to be implemented by

In my discussion of explanatory autonomy I have emphasized that autonomous explanations refer to kinds with distinct determination dimensions. But another element that is necessary is the (tacit) *recognition*, at least by the experts within the linguistic community, that the explanations refer to kinds with distinct determination dimensions. Conversely, explanations that successfully refer to the same level of abstraction are not explanatorily reductive unless there is some *recognition* (if only tacit) of the sameness of determination dimensions for the relevant kinds. This emphasis on recognition is appropriate, because an explanation involves the communication of understanding. How the kinds are understood, at least by some segment of the population (e.g., the relevant experts), is part of the explanatory content. For explanatory autonomy, unlike ontological autonomy, a community's understanding of a kind is relevant to the autonomy of explanations invoking that kind term.

It is also interesting to consider whether ontological difference of determination dimensions is necessary for weak explanatory autonomy, or if ontological sameness of determinations is necessary for explanatory reduction. Suppose, for example, that Jack winces whenever he drinks water. Further suppose that *water* is identical to H_2O. Could the following explanations of Jack's wincing behavior possibly be weakly explanatorily autonomous with respect to one another?

4A. Jack drank water.
4B. Jack drank H_2O.

It could happen, in some linguistic community, that not even the experts know that 'water' and 'H_2O' refer to the same kind. We can even assume that the speakers mistakenly associate different determination dimensions with 'water' and 'H_2O,' but otherwise they are normal English speakers in ordinary environments. So could it be that, for those communities ignorant of the fact that *water* is H_2O, 4A and 4B offer autonomous

what I have been describing as their machine language realizations. Such implementation is a distinct step—e.g., the high-level instructions are compiled into instruction codes that are then executed. The presence of this extra step makes this not a true case of realization. No such step is present, for example, for cognitive psychological kinds. The realization of a cognitive psychological kind, as in all cases of true realization, is synchronic with the psychological state it realizes. One way of making this point is that you cannot have a cognitive psychological kind that fails to be "compiled." All instances of cognitive psychological kinds are necessarily "compiled"—i.e., realized. But a high-level instruction can exist prior to its machine language compilations, and it can also fail to be compiled.

explanations even though, by supposition, they are picking out the very same level of abstraction? We could also construct a parallel case in which the explanations refer to distinct kinds, though the speakers understand them as being the same, and ask if such explanations are the same.

I do not have a vested interest in defending a particular position here. But I am inclined to think that explanatory autonomy requires real distinctness of abstraction and explanatory reduction requires real sameness of abstraction. This is because to count as an explanation at all there must be an adequate fit between the understanding and the world, and this fit is lacking if the same level of abstraction is conceived as distinct or if different levels of abstraction are conceived as the same. It is not necessary to argue for this point, though. For whether or not this is the preferred way to understand explanation, it is clear that determination dimensions are central to settling questions of explanatory autonomy and reduction. Explanatory autonomy requires either the distinctness of accepted (recognized, conceptualized, etc.) determination dimensions *and* the distinctness of actual determination dimensions, or it requires simply the former. And explanatory reduction requires either the sameness of accepted (recognized, conceptualized, etc.) determination dimensions *and* the sameness of actual determination dimensions, or it requires simply the former.

Though they do not develop, and very well might not accept, the theory of kinds offered here, several prominent philosophers of mind have promulgated the view that MR entails weak explanatory autonomy. Daniel Dennett, for example, has claimed that the patterns picked out by the kind terms for multiply realized kinds are distinct from the patterns picked out by the kind terms for their realization kinds. Using a particularly vivid thought experiment, he has claimed that Martians would be missing out on something if the only explanations of human behavior that they understood invoked patterns of the latter kind.

> But if the Martians do not see that indefinitely *different* patterns of finger motions and vocal cord vibrations—even the motions of indefinitely many different individuals—could have been substituted for the actual particulars without perturbing the subsequent operation of the market, then they have failed to see a real pattern in the world they are observing.[30]

[30] Dennett (1987: 26).

In other words, they have failed to understand an explanation that is autonomous with respect to the explanation that they do understand. And Dennett's talk of "indefinitely *different* patterns" is clearly referring to the MR of the behavior in question. Similar claims of weak explanatory autonomy, from considerations of MR alone, can be found in Fodor (1974, 1997) and Pylyshyn (1980).

I have argued that MR entails weak explanatory autonomy, one form of non-reductionism. This conception of MR is consistent with other reductionistic explanatory agendas, however. Multiply realized kinds nevertheless stand in synchronic relations (e.g., realization!) to the lower-level kinds that realize them. Discounting any strong type of emergence, such relations certainly obtain for all higher-level kinds. Given that multiply realized kinds stand in relations both to other kinds within their conceptual scheme and also to the conceptual schemes of their realizations, we can distinguish two different types of explanatory projects. First, there is the search for intra-theoretic explanations which, as has just been argued, operate autonomously. Second, there is the search for inter-theoretic explanations that discover connections between sciences, perhaps investigating necessitation relations in particular. This latter project is certainly reductionistic in some sense. But it is also completely compatible with the weak explanatory autonomy argued for here. These two projects are not intended to exhaust the options, and they certainly are not mutually exclusive. Rather, they are two types of scientific agendas that are welcomed by the current picture of MR. To the extent that multiply realized kinds are blind to realization differences, autonomous intra-theoretic explanations are possible. But to the extent that realization relations hold between lower-level kinds and the higher-level kinds to which they give rise, the inter-theoretic project is viable as well.

Recall that if explanations essentially invoking kind terms for multiply realized kinds are strongly explanatorily autonomous, then they explain things that cannot be explained by any lower-level conceptual scheme. One might wonder whether Putnam, Fodor, Dennett, and Pylyshyn also endorse strong explanatory autonomy when they refer to explanations or patterns that cannot be "seen" from the lower-level science. It is worth noting that in such cases the *explanandum* itself, in addition to the explanation, typically invokes a kind term for a multiply realizable kind (e.g., 'wincing behavior'). Theorists advocating strong explanatory autonomy can then acknowledge that there are lower-level explanations of the bodily movements that constituted Jack's wincing behavior on this particular occasion. However, that lower-level explanation is sensitive to too much information that is irrelevant to its being wincing behavior, and this overspecificity counts against it being an explanation of the wincing. This, I believe, is one of the morals of the stories about Putnam's square peg and Dennett's Martian. Sober (1999) has offered a prominent dissent, arguing that there are lower-level explanations for these higher-level phenomena. To settle this dispute we would have to enter into a more serious discussion of the nature of explanation and the prominence that should be given to the standards of specificity and generality in particular. I rest content with the conclusion that explanations essentially invoking kind terms for multiply realized kinds, where the difference in the determination dimensions for the realized and realization kinds is at least tacitly acknowledged by the experts, are weakly explanatorily autonomous with respect to explanations in terms of their realizations.

5.4.3 *Weak Methodological Autonomy*

MP1 of our MR Argument for Methodological Autonomy claims that weak methodology autonomy follows from MR. Roughly speaking, weak methodology autonomy is understood as follows: A science *X* is weakly methodologically autonomous with respect to some other science *Y* if productive development of *X*-theories, *X*-explanations, and *X*-classifications can proceed in ignorance of *Y*-theories, *Y*-explanations, and *Y*-classifications. Methodological autonomy has many practical consequences associated with teaching and research in that field—e.g., distinctive methodologies to learn and teach; special equipment, such as tools and instruments, to purchase to investigate its kinds; independence from the methodologies and equipment developed for the investigation of its realization kinds, etc. For example, if cognitive psychology is weakly methodologically autonomous with respect to neuroscience, then fruitful development of cognitive psychological theories can proceed in ignorance of neuroscientific theories, explanations, classifications, and methods.

This methodological claim very well might be the most important of our three claims concerning autonomy. This is because it aims to give practical advice as to how science can or should proceed, mainly on the basis of metaphysical considerations alone. For this reason it is also the claim that should be treated with the greatest skepticism and care. Our account of MR suggests that to the extent that realization details are irrelevant to individuating or discovering the essence of multiply realizable kinds, one can have complete understanding of a multiply realizable kind in itself and as it relates to its conceptual scheme—that is, complete understanding of its essence and intra-theoretic connections—while remaining ignorant of realization details. In this sense it is possible *in principle* for theorizing about a multiply realizable kind to be comprehensive and correct, *with respect to the theory it appears in*, even while remaining ignorant of the sciences of its realizations. But this does not mean that attention to realization details will not aid in studying the multiply realizable kinds *in practice*. And, of course, our methodological principles should be governed by what works in practice!

At this point Kim's Challenge, from §4.4.3, is again relevant. If it turns out that there are interesting laws, generalizations, or explanations concerning multiply realized kinds, then these should be discoverable by one who has knowledge of those multiply realized kinds and the other kinds, presumably multiply realized as well, that belong to their conceptual schemes. After all, such generalizations could be confirmed or disconfirmed simply by observing the occurrence or non-occurrence of the relevant multiply realized kinds in appropriate experimental settings. So if Fodor is correct in his response to Kim's Challenge (as I think that he is), then it immediately follows that these sciences of multiply realized kinds are weakly methodologically autonomous.

It would be a mistake to further conclude, at least without some serious argument, that sciences of multiply realized kinds cannot benefit from, and should therefore having nothing to do with, the sciences of their realization kinds. That is, strong methodological autonomy does not follow from weak methodological autonomy. It is an empirical question whether investigation at the realization level can prompt us to discover higher-level generalizations and features that abstract away from this very realization detail. Indeed, the present account of MR even seems compatible with the reductionist methodological strategy recommended by Patricia Churchland (1986), whereby sciences typically coevolve. On her picture, the higher-level kinds are "susceptible to correction and reconceptualization at the behest of the cohort theory."[31] On the present account MR is compatible with this "level jumping" as a methodological

[31] Churchland (1986: 373).

strategy for uncovering the proper higher-level kinds. However, we should reject her more ontological claims that the discovery of realization differences predicts a "fragmentation" of the multiply realized kind. Emphatically, we should still insist that there can be higher-level sameness through lower-level difference. Churchland is correct, however, that ontological reduction is often an epistemic possibility, and this provides reason enough for inter-level investigation. Furthermore, there might be determinates of a higher-level kind that are unlikely to be discovered, as a matter of practical fact, without descending to lower-level realization detail.

5.5 Concepts, Kinds, and Inter-Theoretic Reduction

5.5.1 *Concepts and Kinds*

It might seem that autonomy comes too easily on the present view. I have argued that kinds that differ in their determination dimensions are weakly autonomous in the ontological, explanatory, and methodological senses. But one might point to various examples from the history of science in which reductions of each of these three types have been achieved in spite of apparent differences in determination dimensions. One might offer widely accepted identities such as '*heat* is *mean molecular kinetic energy*' or '*water* is H_2O' as examples. Assuming that these identity statements are true, they pose a problem for the present view if *heat* and *mean molecular kinetic energy*, or *water* and H_2O, differ in their determination dimensions. On the present view if these kinds are in fact identical, then they must share the same determination dimensions. However, some might object that these kinds do not share the same determination dimensions. Such objectors are confused, I believe, because they mistakenly attribute determination dimensions to terms or concepts rather than to the kinds that they pick out.

Terms and concepts are linguistic and psychological entities, respectively. We often possess descriptive content that we associate with both terms and concepts. And most of us, if not all, associate different descriptive content with the terms or concepts for the kinds referred to in the above examples of theoretical identity statements. For example, most of us associate different descriptive content with the terms 'water' and 'H_2O.'

This is true not only of the typical person from centuries ago (if they even possessed the latter concept), but it is also true of the typical person today. It is likely even true of your typical contemporary chemist. One *could* think that the descriptive content associated with such kind terms and kind concepts provides the determination dimensions for the kinds they pick out. If so, then kind terms or kind concepts differing in their descriptive content pick out kinds with different determination dimensions. This would have the consequence that inter-theoretic reduction is impossible. This is because kind terms or concepts from distinct theories will inevitably differ in descriptive content and, on the present suppositions, thereby pick out distinct kinds. This is an unpleasant conclusion, as inter-theoretic reduction surely has occurred. What has gone wrong?

I simply deny that the determination dimensions of a kind are provided by the descriptive content associated with the terms or concepts corresponding to that kind. If they were, then discovering the essence of a kind would simply be a matter of *a priori* reflection on our terms or concepts. The alternative I favor is now a commonplace, after the pioneering work of Saul Kripke and Hilary Putnam.[32] The kind picked out by one of our referring kind terms is, in typical scientific cases at least, not wholly determined by the descriptions that we, or even the experts with respect to that kind term, associate with it. There are real essences, in addition to nominal essences. In typical cases the mind-independent world makes an ineliminable contribution to determining the reference of our kind terms. This is the widely accepted lesson of "Twin Earth" stories. Of course there must be some—typically vague—degree of fit between the world and our understanding of these kinds, or else an eliminative view is to be preferred. A child raised to believe in Santa Claus eventually realizes that Santa Claus does not exist, not that their mother *is* Santa Claus.

Back in Chapter 2, I endorsed, but did not argue for, Property Realism. Though this view is not demanded by the insights of Kripke and Putnam, it does fit nicely with those observations. Let us think of properties as "out there" in the world, with an ontological structure that is exhausted by their determination dimension values and non-determinable necessities. Our terms and concepts somehow latch onto these properties and the kinds to which they give rise. But our terms and concepts typically do not provide us with complete knowledge of these properties and their kinds.

[32] Putnam (1975) and Kripke (1980).

And sometimes our terms and concepts latch onto kinds even though the descriptive content that we associate with them is not even true of these kinds. Of course, it is well known that similar points hold for ordinary names. If the assumption that properties and their determination dimensions are "out there" to be discovered is correct, then an account of kind-level ontological reduction is at hand once we develop an account of how kind terms latch onto their kinds and we determine the degree and type of fit between descriptive content and world required for successful kind reference to obtain. For scientific kind terms and kind concepts this descriptive content will, in large part, be provided by the role the term or concept occupies in the laws, generalizations, and explanations of that science.

5.5.2 *Examples: Chemical Reductions*

There are many scientific examples of reductions that were driven by discovering what might originally be taken to be the single realization of some kind. Chemistry, for example, has discovered that there are singular chemical realizations for some of our folk kinds. These realization claims were ultimately converted into identity claims. Here are three examples:

Folk Classification	**Chemical Realization**
Water	H_2O
Gold	*Au*
Salt	*NaCl*

Lay people, scientists, and professional philosophers agree, on the whole, that *water* is identical to H_2O, *gold* is identical to *Au*, and *salt* is identical to *NaCl*.[33] It seems as if the folk and the philosophers *infer* these identities from the fact of SR.[34] It is because we find only one chemical element associated with gold that we identify *gold* with *Au*. It is because we find only one

[33] This is supposed to be true on Kripke–Putnam theories of reference for natural kind terms, according to which they are rigid designators that refer to certain kinds with which the community of speakers is causally connected. See Putnam (1975) and Kripke (1980).

[34] Strictly speaking, SR is not required for identity—only *dominant* realizability in the sample population is required. In cases of dominant realizability the reduction is made with a slight modification in our classifications.

chemical substance associated with water that we identify *water* with H_2O. Significantly, those few who deny that *water* is identical to H_2O likely hold that *water* is multiply realizable—say, by *XYZ*, H_2O, etc. It would be bizarre to hold that *water* is singularly realizable by H_2O, but nevertheless *water* and H_2O are distinct kinds at different levels of abstraction. Chemistry provides us with a host of examples in which SR appears to entail kind identity.

These are all examples in which a word from everyday English supposedly picks out a single chemical kind. Most agree that the kinds picked out by these everyday English words are identical to the kinds picked out by the more technical vocabulary of chemistry. Of course identity is a one-to-one relation involving the very same thing, but it is natural to call such ontological reductions folk-*to*-scientific reductions given that the scientific concept better matches the kind and its instances—in the sense of containing information more pertinent to the essence of that kind and its instances—than does our folk concept. 'Water' and 'H_2O' pick out the very same kind, but only the latter is in the vocabulary of a science.

One might accept that *water* is identical to H_2O but deny that this identity holds in virtue of water being singularly realized by H_2O. Rather, this objector might claim, *water* is identical to H_2O because all along the concept of water was understood, either implicitly or explicitly, as referring (if it referred at all) to a chemical kind. Though the nature of the chemical kind *water* was unknown, it was always understood to be some chemical kind or other. On this view the discovery of SR did not establish that *water* is a chemical kind. That was known all along. Rather, that discovery identified the chemical kind to which *water* is identical.

But this is implausible. It is true that, among some of the intellectual elites at least, 'water' was thought to refer to a chemical kind. Among such intellectuals it was believed, for at least a couple thousand years, that water is a physical element. But it is worth considering how they would have reacted had they discovered, contrary to actual fact, that *water* is neither a physical element nor any other chemical kind. Imagine a world just like ours except that various chemical kinds—*XYZ*, H_2O, and others—are well-represented in the samples speakers intend to refer to with the term 'water.'[35] How would, and how should, a community of speakers from this world react upon discovering that there are many chemical kinds that they are gesturing at with their general term 'water'? I think it is very unlikely

[35] Compare this discussion with Soames (2002: ch. 10).

that, on the basis that there is this variety in chemical constitution, they would conclude that *water* does not exist. I also think that it would be absurd for them to do so. They need not conclude that there is no such kind as *water* or even that *water* is not a natural kind. Instead, I take it that they would, and should, simply conclude that, contrary to their expectations, *water* is a kind that is multiply realized in the chemical. And perhaps *water* would still be a natural kind, as there might be multiply realized natural kinds. (An interesting question here is whether this community of speakers would, and should, allow that there could be other chemical kinds, besides those found on "their Earth," that realize *water*.)

In this example the speakers intended or expected to refer to a single chemical kind, but they were actually referring to something realized by a variety of chemical kinds. Nevertheless, they successfully referred to (their) *water*. Now consider a different counterfactual situation. Imagine that a community of speakers explicitly did not intend to refer to a single chemical kind in their use of 'water,' though they took themselves to be picking out, at least superficially, the stuff occupying the familiar water role. That is, they used the word 'water' to refer to the stuff they drink that is found in lakes and oceans, etc., but they explicitly did not assume nor demand that this stuff is chemically homogeneous. In fact, we could even tell the story such that they explicitly believed that *water* is realized by many diverse chemical kinds, perhaps because they had discovered that their other folk kinds were multiply realized in the chemical. It is worth considering how such a community would respond once they discovered that there is in fact only one chemical kind—H_2O—that realizes *water* on their planet. Contrary to their intentions, they might conclude that *water* has only one possible chemical realization, H_2O, and that it is identical to H_2O. Alternatively, they could retain their belief that *water* is multiply realizable and simply conclude that their planet happens to contain only one of these realization kinds. I assume, without any argument but for an appeal to our actual practices, that the former outcome is more likely.

If the claims made about these two counterfactual situations are correct, then it is not the intention to refer to a chemical kind that occupies a certain role in our daily life, in conjunction with the world being such that the chemical kind H_2O occupies this role, that secures the reduction of *water* to H_2O. Rather, this reduction is secured by the simple fact that there is a dominant chemical realization on Earth that fits the superficial characterization of water. The intention to refer to a chemical kind is irrelevant to the reduction, because the reduction would hold whether or not the intention to refer to a chemical kind is present. The first example is supposed to show that if this intention were actually present but the stuff meeting the superficial characterization of water were chemically heterogeneous, water would still exist. In such a situation, contrary to the user's intentions, they would not be referring to a chemical kind. Nevertheless, they would be successfully referring to (their) *water*. And the second example is supposed to show that if the language users explicitly rejected any intention of referring to a single chemical kind, they might nevertheless be doing so should their planet be such that *water* has a singular realization.

Of course, a community of speakers could use the word 'water' such that they *demand*, rather than merely intend or expect, that water be singly realized. Alternatively, they might demand that water be multiply realized, if 'water' is to refer at all. If such people insisted on sticking to their demands in spite of the discovery that the world does not meet them, then the reduction would be thwarted and (their) water would not exist. Still, it is overwhelmingly likely that such people would coin a new term for the wet stuff that surrounds them were they to discover that, contrary to their previous demands, the wet stuff is multiply realized or, in the alternate case, singularly realized.

These claims are supported by the fact that in the actual world 'water,' at least among the intellectual elite, was widely thought to refer to a physical element. It was not until the 1780s that Lavoisier, Watt, and others made the chemical discoveries that proved water is a chemical compound rather than an element. I am aware of no movement that, upon this discovery, claimed that water does not exist because there is no such element. We easily moved from conceiving of a substance as an element to conceiving of it as a compound. Though comparisons here are admittedly difficult, it seems that the shift to conceiving of something as a compound rather than as an element is similar to the shift to conceiving of something as realized by multiple chemical compounds rather than being realized by a single chemical compound. If we retain our belief in the existence of water in light of the discovery that this wet stuff is a compound rather than an element, so too should we retain our belief in the existence of water in light of the discovery that this wet stuff is realized by multiple chemical compounds rather than by a single chemical compound.

These examples display an interesting fact about kinds. Mere singular realization, a non-modal concept, is sometimes sufficient for SR. This is one of the lessons that we should draw from Kripke/Putnam-style examples. One of the first lessons we learned from Twin Earth type examples is that our kind terms have an indexical component and they refer to things with essences largely determined by factors that are not in our heads. If Kripke and Putnam are right, then the world itself largely determines these essences rather than these essences being determined by the intentions or descriptions we associate with those kind terms. This much can be accepted, however, without bestowing a privileged role on chemistry or any other science when it comes to determining the essences of kinds. My addition to this story, as presented through the above discussion, is that whether the essence of a particular kind is provided by, say, its chemical constitution as opposed to being provided by some higher or lower (or simply different) level of abstraction often depends on whether the dominant samples of that kind are singularly or multiply realized by chemical kinds. In the next section I further develop this role for SR.

5.5.3 *Singular Realizability*

There are complicated and interesting connections between the single realization, SR, and identity of kinds. As we just discussed, the empirical discovery of the singular realization of a kind is sometimes taken to be sufficient to establish that a kind term refers to a kind that does not have other metaphysically possible realizations (i.e., it is singularly realiz*able*) and that the kind picked out by this kind term is identical to this singular realization kind. This seems to be the accepted view of water. There is in fact one chemical realization of water—H_2O. There are no other metaphysically possible realizations of water. And, water is identical to H_2O. But these are three distinct claims. The first was an empirical discovery. It seems that the next two claims—SR and kind identity—are then inferred from the fact of singular realization.

Scott Soames has reconstructed different versions of an argument, inspired by Kripke's Lecture III from *Naming and Necessity*, which concludes that the kinds picked out by coextensive kind terms of the same "type" (e.g., substance terms, species terms, etc.) are identical. Soames provides a few formalizations of such arguments, but here is a good summary of the line of thought:

> On Kripke's account, the linguistic mechanism that fixes the extension of an ostensive natural kind predicate guarantees that if it has successfully been introduced, then there can be only one natural kind, of the specified type T [e.g., substance, species, etc.], of which most members of the sample are instances. Thus, when A is ostensive, and the predicate B is related to A in the appropriate way, finding

out that everything in the extension of A is in the extension of B is tantamount to finding out that the kind associated with A is identical with the kind associated with B. This powerful conclusion does not come from nowhere. Rather, it comes from the fact that the semantic presuppositions associated with the introduction of natural kind predicates make very substantial nonlinguistic claims—including the claim that the individuals in the sample associated with the ostensive predicate A are members of a single natural kind of a given type, and the claim that the predicates A and B designate kinds of the same type.[36]

Though he does not use this terminology, he is considering cases of singular realization. The reasoning here is like that supporting our water example from the previous paragraph, the primary example used by Soames as well. Applying the argument here, 'water' and 'H_2O' are both supposed to be kind terms for substances. Since (almost) all and only the samples of water are also samples of H_2O, *water* is identical to *H_2O*. This quick move to the identity claim obviously cuts off the possibility of MR. Singular realization (i.e., coextensiveness), a non-modal notion, is taken to be sufficient for SR, a modal notion.

This argument comes from the philosophy of language. Perhaps Soames is correct that, at least for most of our kind terms, such coextensiveness is sufficient to establish that the kind terms refer to the same kind. This is even more plausible if we limit ourselves to the kind terms Soames focuses on, substance terms and species terms, and accept some widely endorsed metaphysical assumptions—namely, that substances are physical constitution kinds and explanatory kinds. Assume that 'water' is a substance term. The physical constitution of water is H_2O, and the observable properties of water can be explained by this chemical constitution. So, *water* is *H_2O*. But we should also challenge these restrictions and assumptions to see if this form of argument generalizes to all cases of coextensiveness (or, in my terminology, singular realization).

We could certainly challenge the claim that substantival kind terms must pick out kinds individuated by their physical constitution (e.g., chemical makeup) or by the underlying properties that explain their observable features. Though sympathetic to these moves, I will neither push nor develop them here. Rather, I want to point out that the Soames–Kripke argument is less plausible when applied to adjectival kind terms like those to which we have primarily attended. If there are multiply

[36] Soames (2002: 269).

realizable kinds at all, then it should be obvious that singular realization does not entail identity. Suppose that *pain* is a multiply realizable kind. There are nevertheless worlds in which *pain* has, as a matter of fact, only one kind of realization—say, NS_1. We conceive of such a world by simply "subtracting away" all the other possible realizations. It might be that, in such a world, their word 'pain' (assuming that they speak like our English speakers) refers to NS_1. Let's pretend that this is how the reference of kind terms works. But, as a metaphysical point, it would nevertheless be true that in this singular realization world (our) *pain* is still a kind distinct from NS_1, with distinct and actual instances as well. It might even be possible for inhabitants of such a world to recognize that there are other metaphysically possible realizations of *pain* in addition to NS_1. And they could recognize the distinct determination dimensions, and hence essences, of the pains and neuroscientific states in their world. The inference from singular realization to identity does not work for adjectival kinds.

The inference to identity has much greater plausibility, however, when we move from single realization to apparent SR.[37] I have no knockdown argument to support this claim, but there are advantages to accepting that SR (the modal notion) very likely entails an identity. We are considering a situation in which instances of kind *B* metaphysically necessitate instances of kind *A* and are co-located with them. Further, given the supposed SR, we are supposing that instances of kind *A* at least nomologically necessitate instances of kind *B*. Imagine that $pain_{1492}$ were singularly realizable by NS_1 in this sense. In this case there is significant pressure to say that $pain_{1492}$ is identical to NS_1, and so too for any other case of apparent SR.

First, remember that instances of scientific kinds are out there in the world with essences to be empirically discovered. Even if Property Realism is not true, we should still be guided by the same standards for discovering determination dimensions that the scientifically astute Property Realist would accept—explanatory power, simplicity, coherence, etc. The Property Realist and Property Nominalist can both equally accept or deny that, say, hue, brightness, and saturation are the determination dimensions of *color*. And they could do so for the same empirical and theoretical

[37] I use the qualification 'apparent' because if identity holds then, strictly speaking, it would be a misuse to speak of this as singular *realizability*. For, a kind cannot realize itself. Nevertheless, I will often, speaking loosely, write of SR sometimes entailing identity. The reader should remember that, strictly speaking, such cases are not cases of realizability at all.

reasons—their metaphysics of properties is irrelevant here. The essence of an adjectival kind is exhausted by its determination dimension values and non-determinable necessities, and for scientific kinds these are largely determined on empirical grounds. For our imagined situation the critical question is whether $pain_{1492}$ and NS_1 have the same determination dimensions and non-determinable necessities. If so, they are identical. If not, they are distinct and we have real SR (without identity). One reason to go with the identity option is that it better coheres with the distinction between our conception of a kind and its mind-independent essence. Given the mind independence of properties and their determination dimensions, the same property or determination dimension could be conceived in different ways. This may be the best explanation of the situation in our imagined case. When thought of under the pain_{1492}-concept, we think of the property as a phenomenological intensity. But, when thought of under the NS_1-concept, we think of the property as a neuroscientific structure. These could be two ways of conceiving of the very same property, in which case an ontological reduction is in order.

There are many possible complications to the situation, however. Suppose we originally associated one determination dimension with $pain_{1492}$ but discover that its singular realization kind NS_1 has a complex five-dimensional property space. We could pursue the reduction option by modifying our understanding of $pain_{1492}$ to meet the presumably more nuanced and scientifically profitable understanding of NS_1 as a region within a five-dimensional property space. We would then accept the kind identity. Other times it is simply too much of stretch to modify the understanding of a concept to secure an identity, and an elimination is in order. Such might be the case with our refusal to identify *dephlogisticated air* with *oxygen*. Instead, we just reject the kinds posited by the phlogiston theory.

Another complication concerns whether other kinds belonging to the super-determinable property space *pain* are multiply or singularly realizable. Could it be that though $pain_{1492}$ is singularly realizable, $pain_{1066}$, $pain_{1776}$, and various other *pain* kinds are multiply realizable? Such a situation would thwart efforts to identify $pain_{1492}$ with NS_1. This is because such an identity would require unifying their determination dimensions, though the MR of other *pain* kinds would show that the determination dimension(s) of *pain* is distinct from those of the *NS*-kinds. (This point is reminiscent of our response, in §5.2, to Haug's objection to the present account of MR.) But surely $pain_{1492}$ is a determinate of *pain*, and thus it shares its determination dimension(s). If both the *pain* kinds and *NS*-kinds are well entrenched in theories that offer fruitful explanations of their phenomena, then in such situations we do have reason to allow for the distinctness of a kind, $pain_{1492}$, and its only possible realization, NS_1. However unlikely such a situation might be, such situations would seem to be legitimate cases of SR.

Absent that kind of situation, however, SR should push us towards identification. First, it is always possible to have multiple conceptions of the same property or kind. With cases of SR we should suspect that this might be the case, as identity offers an explanation of the necessary co-occurrence of such properties. Second, unification is an explanatory virtue and we should see if one or the other classificatory system has a significant advantage over the other. If so, and depending on the relations that other kinds belonging to their super-determinable property spaces bear to one another, elimination or modification (e.g., of the determination dimensions associated with a kind) are possible options.

5.6 Concluding Thoughts

Properties are abstract and specific. This is a theme that has been emphasized throughout this book: the distinction between abstraction and specificity. Levels of abstraction are given by determination dimensions and non-determinable necessities. Specificity is given by values, or ranges of values, along determination dimensions. Since properties are particular, they are assumed to be super-determinate. Generality exists only with kinds, which can be determinable. But generality ought not to be confused with abstraction. Super-determinate kinds are still abstract, even while maximally specific.

I have aimed to develop and defend a property theory that explains identity, determination, and realization. The individuation conditions for properties are derivative from an understanding of determination. Namely, determination dimension values and non-determinable necessities provide us with the essence of a property. Determinable kinds and determinate kinds are at the same level of abstraction, differing only in their specificity. Realizer and realized kinds, however, are at different levels of abstraction. As they do not share determination dimensions, realizer kinds cannot specify the kinds that they realize (contrary to, say, subset of causal powers accounts of realization). For this very same reason, these kinds cannot share an instance (property). So we see that the distinction between abstraction and specificity neatly accounts for the difference between determination and realization. Determinable kinds and determinate kinds have sameness of abstraction, and differ only in specificity; realizer kinds and realized kinds differ in abstraction, voiding specificity comparisons.

Throughout this book, I have stated my conviction that properties *really* exist. In large part this is due to the role they play in causation. If this is correct, then the essences of properties can be discovered by investigating the essences of causes and effects. This seems quite reasonable to me, but pursuing this line would be a project of its own. Properties are out there in the world, and it is a mistake to think that their essences—these determination dimensions I have written so much about—are the products of our minds. They are out there in the world to be discovered. There is a sense in which this is true for the Nominalist, as well. The Realist has certain procedures and standards for positing determination dimensions, and these are guided by the data collected and the norms of good investigation as accepted within the relevant conceptual scheme (scientific methodology is particularly well-established). But the data and norms that shape our conceptions of the kinds posited by the various conceptual schemes should guide us just the same, regardless of whether we accept Realism or Nominalism about properties. Neither party thinks that our conceptions are infallible, and both parties (for the most part) believe that our conceptions should fit the structure of the world. And, as I have argued as the central claim of this book, all such kinds have a logical structure which comprises only determination dimensions and non-determinable necessities. This is the sense in which kinds and properties are abstractions.

Bibliography

Aristotle. 1984. *The Complete Works of Aristotle*, ed. Jonathan Barnes. Princeton, NJ: Princeton University Press.

Armstrong, David. 1978. *Universals and Scientific Realism*. New York, NY: Cambridge University Press.

Armstrong, David. 1997. *A World of States of Affairs*. New York, NY: Cambridge University Press.

Atkinson, R. C. and R. M. Shiffrin. 1968. "Human Memory: A Proposed System and its Control Processes," in K. W. Spence and J. T. Spence (eds), *The Psychology of Learning and Motivation*, Volume 2. New York, NY: Academic Press, 89–195.

Baddeley, A. D. 1966. "The Influence of Acoustic and Semantic Similarities on Long-Term Memory for Word Sequences," *Quarterly Journal of Experimental Psychology*, 18: 302–309.

Baddeley, A. D. and G. J. Hitch. 1974. "Working Memory," in G. H. Bower (ed.), *The Psychology of Learning and Motivation*, Volume 8. New York, NY: Academic Press, 47–89.

Bechtel, William and Jennifer Mundale. 1999. "Multiple Realizability Revisited: Linking Cognitive and Neural States," *Philosophy of Science*, 66: 175–207.

Bickle, John. 1998. *Psychoneural Reduction: The New Wave*. Cambridge, MA: MIT Press.

Bickle, John. 2003. *Philosophy and Neuroscience: A Ruthlessly Reductive Account*. Norwell, MA: Kluwer Academic Publishers.

Bjork, Elizabeth and Robert Bjork. 1996. *Memory*. San Diego, CA: Academic Press.

Block, Ned. 1997. "Anti-Reductionism Slaps Back," *Philosophical Perspectives*, 11: 107–132.

Blum, Richard. 2005. *Professional Assembly Language*. Indianapolis, IN: Wiley Publishing.

Boyd, Richard. 1999. "Kinds, Complexity and Multiple Realization," *Philosophical Studies*, 95: 67–98.

Churchland, Patricia. 1986. *Neurophilosophy*. Cambridge, MA: MIT Press.

Chalmers, David. 1996. *The Conscious Mind*. New York, NY: Oxford University Press.

Chalmers, David. 2010. *The Character of Consciousness*. New York, NY: Oxford University Press.

Clapp, Lenny. 2001. "Disjunctive Properties: Multiple Realizations," *Journal of Philosophy*, 98: 111–136.

Crane, Tim. 2008. "Causation and Determinable Properties: On the Efficacy of Colour, Shape and Size," in Jakob Hohwy and Jesper Kallestrup (eds), *Being Reduced: New Essays on Reduction, Explanation, and Causation*. New York, NY: Oxford University Press, 176–195.

Craver, Carl. 2001. "Role Functions, Mechanisms, and Hierarchy," *Philosophy of Science*, 68/1: 53–74.

Cummins, Robert. 1983. *The Nature of Psychological Explanation*. Cambridge, MA: MIT Press.

Davidson, Donald. 1980. *Essays on Actions and Events*. New York, NY: Oxford University Press.

Dennett, Daniel. 1987. *The Intentional Stance*. Cambridge, MA: MIT Press.

Dretske, Fred. 1988. *Explaining Behavior: Reasons in a World of Causes*. Cambridge, MA: MIT Press.

Ehring, Douglas. 1996. "Mental Causation, Determinables, and Property Instances," *Noûs*, 30/4: 461–480.

Ehring, Douglas. 2003. "Part-Whole Physicalism and Mental Causation," *Synthese*, 136/3: 359–388.

Ehring, Douglas. 2011. *Tropes: Properties, Objects, and Mental Causation*. New York, NY: Oxford University Press.

Feigl, Herbert. 1958. "The 'Mental' and the 'Physical,'" *Minnesota Studies in the Philosophy of Science*, 2: 370–497.

Fine, Kit. 1994. "Essence and Modality," *Philosophical Perspectives*, 8: 1–16.

Fodor, Jerry. 1974. "Special Sciences, or the Disunity of Science as a Working Hypothesis," *Synthese*, 28: 97–115.

Fodor, Jerry. 1997. "Special Sciences: Still Autonomous After These Years," *Philosophical Perspectives*, 11: 149–163.

Funkhouser, Eric. 2006. "The Determinable-Determinate Relation," *Noûs*, 40/3: 548–569.

Funkhouser, Eric. 2007a. "A Liberal Conception of Multiple Realizability," *Philosophical Studies*, 132/3: 467–494.

Funkhouser, Eric. 2007b. "Multiple Realizability," *Philosophy Compass*, 2/2: 303–315.

Gaines, Brian. 2009. "Designing Visual Languages for Description Logics," *Journal of Logic, Language and Information*, 18/2: 217–250.

Gillett, Carl. 2002. "The Dimensions of Realization: A Critique of the Standard View," *Analysis*, 62/4: 316–323.

Gillett, Carl. 2003. "The Metaphysics of Realization, Multiple Realizability, and the Special Sciences," *Journal of Philosophy*, 100: 591–603.

Gillett, Carl and Bradley Rives. 2005. "The Non-Existence of Determinables: Or, a World of Absolute Determinates as Default Hypothesis," *Noûs*, 39/3: 483–504.

Haug, Matthew. 2010. "Realization, Determination, and Mechanisms," *Philosophical Studies*, 150/3: 313–330.

Heil, John. 1999. "Multiple Realizability," *American Philosophical Quarterly*, 36/3: 189–208.

Heil, John. 2005. *From an Ontological Point of View*. New York, NY: Oxford University Press.

Heil, John and C. B. Martin. 1999. "The Ontological Turn," *Midwest Studies in Philosophy*, 23: 34–60.

Hooker, Clifford. 1981. "Towards a General Theory of Reduction (Parts I–III)," *Dialogue*, 20: 38–59, 201–236, and 496–529.

Johnson, W. E. 1921. *Logic*, Part I. London: Cambridge University Press.

Kim, Jaegwon. 1993a. *Supervenience and Mind*. New York, NY: Cambridge University Press.

Kim, Jaegwon. 1993b. "Multiple Realization and the Metaphysics of Reduction," reprinted in Kim 1993a, 309–335.

Kim, Jaegwon. 1998. *Mind in a Physical World*. Cambridge, MA: MIT Press.

Kim, Jaegwon. 2005. *Physicalism, or Something Near Enough*. Princeton, NJ: Princeton University Press.

Korner, Stephan. 1959. "On Determinables and Resemblance, I," *Proceedings of the Aristotelian Society Supplementary*, 33: 125–140.

Kripke, Saul. 1980. *Naming and Necessity*. Cambridge, MA: Harvard University Press.

Lewis, David. 1983. *Philosophical Papers*, Volume 1. New York, NY: Oxford University Press.

Lewis, David. 1986. *Philosophical Papers*, Volume II. New York, NY: Oxford University Press.

Lewis, David. 1999. *Papers in Metaphysics and Epistemology*. New York, NY: Cambridge University Press.

Locke, John. 1975. *An Essay Concerning Human Understanding*, ed. Peter H. Nidditch. Oxford: Clarendon Press.

Lycan, William. 1981. "Form, Function, and Feel," *The Journal of Philosophy*, 78: 24–50.

Lycan, William. 1990. "The Continuity of Levels of Nature," in William Lycan (ed.), *Mind and Cognition*. Malden, MA: Blackwell Publishing, 77–96.

MacBride, Fraser. 2005. "The Particular–Universal Distinction: A Dogma of Metaphysics?" *Mind*, 114/455: 565–614.

Macdonald, Cynthia and Graham Macdonald. 1986. "Mental Causes and Explanation of Action," in L. Stevenson, R. Squires, and J. Haldane (eds), *Mind, Causation, and Action*. Oxford: Basil Blackwell, 35–48.

Miller, George. 1956. "The Magical Number Seven," *The Psychological Review*, 63: 81–97.

Millikan, Ruth Garrett. 1999. "Historical Kinds and the 'Special Sciences,'" *Philosophical Studies*, 95: 45–65.

Place, U. T. 1956. "Is Consciousness a Brain Process?" *British Journal of Psychology*, 47: 44–50.

Plato. 1997. *Plato: Complete Works*, ed. John M. Cooper. Indianapolis, IN: Hackett Publishing.

Polger, Thomas. 2002. "Putnam's Intuition," *Philosophical Studies*, 109: 143–170.

Prior, Arthur. 1949. "I. Determinables, Determinates, and Determinants," *Mind*, 58/229: 1–20.

Putnam, Hilary. 1975. *Mind, Language and Reality: Philosophical Papers*, Volume 2. New York, NY: Cambridge University Press.

Pylyshyn, Zenon. 1980. "Computation and Cognition: Issues in the Foundation of Cognitive Science," *Behavioral and Brain Sciences*, 3/1: 111–134.

Quine, W. V. 1969. *Ontological Relativity and Other Essays*. New York, NY: Columbia University Press.

Richardson, Robert. 1979. "Functionalism and Reductionism," *Philosophy of Science*, 46: 533–558.

Robb, David. 1997. "The Properties of Mental Causation," *Philosophical Quarterly*, 47/187: 178–194.

Rosen, Gideon. 2010. "Metaphysical Dependence: Grounding and Reduction," in Bob Hale and Aviv Hoffmann (eds), *Modality: Metaphysics, Logic, and Epistemology*. Oxford: Oxford University Press, 109–136.

Schaffer, Jonathan. 2003. "The Problem of Free Mass: Must Properties Cluster?" *Philosophy and Phenomenological Research*, 66/1: 125–138.

Searle, John. 1959. "Determinables and the Notion of Resemblance," *Proceedings of the Aristotelian Society*, Supp. Vol. 33: 141–158.

Shapiro, Lawrence. 2000. "Multiple Realizations," *The Journal of Philosophy*, 97/12: 635–654.

Shapiro, Lawrence. 2004. *The Mind Incarnate*. Cambridge, MA: MIT Press.

Shoemaker, Sydney. 2001. "Realization and Mental Causation," in Carl Gillett and Barry Loewer (eds), *Physicalism and Its Discontents*. New York, NY: Cambridge University Press, 74–98.

Shoemaker, Sydney. 2003. *Identity, Cause, and Mind: Philosophical Essays*. New York, NY: Oxford University Press.

Shoemaker, Sydney. 2007. *Physical Realization*. New York, NY: Oxford University Press.

Sider, Theodore. 2003. "What's So Bad about Overdetermination?" *Philosophy and Phenomenological Research*, 67: 719–726.

Sider, Theodore. 2011. *Writing the Book of the World*. New York, NY: Oxford University Press.

Smart, J. J. C. 1959. "Sensations and Brain Processes," *Philosophical Review*, 68: 141–156.

Sober, Elliott. 1999. "The Multiple Realizability Argument Against Reductionism," *Philosophy of Science*, 66: 542–564.

Soames, Scott. 2002. *Beyond Rigidity: The Unfinished Semantic Agenda of Naming and Necessity*. New York, NY: Oxford University Press.

Tanenbaum, Andrew. 1999. *Structured Computer Organization*, 4th edition. Upper Saddle River, NJ: Prentice Hall.

Walter, Sven. 2007. "Determinables, Determinates, and Causal Relevance," *Canadian Journal of Philosophy*, 37/2: 217–243.

Williams, D. C. 1931. "The Nature of Universals and of Abstractions," *The Monist*, 41: 583–593.

Wilson, Jessica. 2009. "Determination, Realization, and Mental Causation," *Philosophical Studies*, 145: 149–169.

Wilson, Jessica. 2012. "Fundamental Determinables," *Philosophers' Imprint*, 12/4: 1–17.

Wilson, Robert. 2001. "Two Views of Realization," *Philosophical Studies*, 104: 1–31.

Yablo, Stephen. 1992. "Mental Causation," *The Philosophical Review*, 101/2: 245–280.

Index

The manufacturer's authorised representative in the EU for product safety is Oxford University Press España S.A. of El Parque Empresarial San Fernando de Henares, Avenida de Castilla, 2 – 28830 Madrid (www.oup.es/en or product.safety@oup.com). OUP España S.A. also acts as importer into Spain of products made by the manufacturer.

Printed in the USA/Agawam, MA
July 30, 2026

908341.029